T0358297

Thermoelectric Energy Conversion Devices and Systems

WSPC Series in Advanced Integration and Packaging

ISSN: 2315-473X

Series Editors: Avram Bar-Cohen *(University of Maryland, USA)*
Shi-Wei Ricky Lee *(Hong Kong University of Science and Technology, ROC)*
Bontae Han *(University of Maryland, USA)*

In the 21st century, innovative integration and packaging drives success in the marketplace, establishing the "look and feel" of electronic products and their price point, and enabling differentiation and branding. The rapid development and proliferation of design methodologies, assembly techniques, and thermal solutions for components ranging from integrated circuits to photonics, and from sensors and MEMS to mixed signals and power amplifiers, drive the often frenetic pace of exploration, discovery, and innovation characteristic of the electronic industry.

This recently introduced WSPC book series on "Advanced Integration and Packaging of Electronic Systems" (AIPack), with a particular focus on 2.5D and 3D form factors, will provide a "technology commons" for the packaging community, a place where ideas can be articulated, examined, and disseminated rapidly and efficiently through both print and electronic media. It is our hope that product developers will find AIPack a convenient and reliable source of technical information; that researchers and students will be guided to the challenges and barriers facing key technologies, and that technology managers and investors will find the tools and data needed to make broad technology assessments and comparisons for emerging packaging and integration technologies.

Published

More information on this series can also be found at https://www.worldscientific.com/series/wssaip

WSPC Series in Advanced Integration and Packaging | Vol. 7

Thermoelectric Energy Conversion Devices and Systems

Kazuaki Yazawa
Purdue University, USA

Je-Hyeong Bahk
University of Cincinnati, USA

Ali Shakouri
Purdue University, USA

World Scientific

NEW JERSEY · LONDON · SINGAPORE · BEIJING · SHANGHAI · HONG KONG · TAIPEI · CHENNAI · TOKYO

Published by

World Scientific Publishing Co. Pte. Ltd.
5 Toh Tuck Link, Singapore 596224
USA office: 27 Warren Street, Suite 401-402, Hackensack, NJ 07601
UK office: 57 Shelton Street, Covent Garden, London WC2H 9HE

Library of Congress Control Number: 2020947143

British Library Cataloguing-in-Publication Data
A catalogue record for this book is available from the British Library.

WSPC Series in Advanced Integration and Packaging — Vol. 7
THERMOELECTRIC ENERGY CONVERSION DEVICES AND SYSTEMS

ISBN 978-981-121-826-2 (hardcover)
ISBN 978-981-121-827-9 (ebook for institutions)
ISBN 978-981-121-828-6 (ebook for individuals)

For any available supplementary material, please visit
https://www.worldscientific.com/worldscibooks/10.1142/11770#t=suppl

Desk Editors: Balasubramanian Shanmugam/Steven Patt

Typeset by Stallion Press
Email: enquiries@stallionpress.com

Printed in Singapore

Preface

Thermal energy enabled industrial revolution and electrical energy allowed major developments in the 20th century. Even today, majority of commercial devices do not directly convert between these two forms of energy but through an intermediate mechanical energy. In electrical power plants, burning of fossil fuels is first converted to steam which moves a turbine. In thermal air conditioning in buildings, consumption of electricity is used for driving compression and expansion of a working fluid. While direct conversion of heat into electricity and electricity into cooling were observed by Seebeck and Peltier almost two centuries ago, solid-state thermoelectric devices have only limited niche applications. There was a major breakthrough after the second world war, when Ioffe and other scientists made the first practical thermoelectric power generators and Peltier coolers. This enabled radioisotope thermoelectric generators for space applications and miniature Peltier coolers for semiconductor lasers, ultra-low noise detectors and other electronic applications. There is a renewed interest in solid-state thermoelectrics since 1990s with the development of nanostructured and complex materials and deeper understanding of the interaction between heat and electricity at the atomic scale.

There are excellent textbooks about fundamentals of thermoelectrics, recent advances in materials, as well as overview of some applications. This book steams from authors' complementary background in electrical and mechanical engineering, thermodynamics, heat transfer, as well as nanostructured materials and quantum mechanics. Our active research is ranging from development of novel thermoelectric materials to thin film devices and applications both in cooling and power generation. Quantitative analysis of the trade-offs in material parameters as well as full system evaluation with realistic electrical and thermal boundary conditions are essential to optimize thermoelectric energy conversion, but rarely discussed in textbooks.

In this book, we explore diverse applications of thermoelectric technologies with state-of-the-art simulations tools. Unique features of thermoelectrics for applications include noise-free, maintenance-free, scalable to small down to microscale, and more. For example, thermoelectric modules are well suited to the on-site power supply demand for the emerging Internet of things (IoT) if heat source exists. We aimed to demonstrate the benefit of analytical modeling, where the system level energy transfer and material properties are equally important for applications and mutually linked in maximizing the energy efficiency and minimizing the cost. Many off-the-shelf thermoelectric modules may work as both generators and coolers with proper electrothermal co-design. Especially for cooling, additional heat rejection is required due to the power consumption in the thermoelectric module, hence an insufficient heat sinking will cause a design failure.

This book covers the coupled electrothermal transport from fundamental to devices and systems. While the intrinsic material thermoelectric figure of merit, ZT, is important, the ultimate performance of the energy conversion system is determined by the dimensions, parasitic resistances and also modified by the external electrical and thermal components. The simulation tools including all system components are powerful to explore the variational analysis and identify trade-offs in efficiency and cost performance.

In the following 10 chapters, diverse topics are covered ranging from basic fundamentals, system integration, application-specific designs, characterization techniques, and some comments about future developments. We hope this book is useful for the readers who are particularly interested in a deeper understanding of thermoelectrics and their system integration.

Chapter 1 briefly describes the discovery of thermoelectricity and introduces the basic elements in this book. A thermoelectric energy conversion device is a thermodynamic system and thus can be compared to other thermodynamic cycles. Subsequently, the role of heat transfer in the irreversible thermodynamic heat engines is explained. Unlike other technologies, the performance of a thermoelectric device is virtually independent of the scale. At the end of this chapter, large-scale impact by integrating thermoelectrics into existing thermal energy systems is also discussed.

Chapter 2 describes how thermoelectrics devices can be used in photonics and electronics applications. Thermoelectric devices are useful for energy-efficient spot cooling, precise temperature control, and for scavenging waste heat. The main advantages are compactness, lightweight, fast response, precise temperature control, and scalability of the thermal solution.

In contrast to the previous chapter, Chapter 3 provides a set of correlations and equations that describe the heat transfer inside and outside of the thermoelectric device and at the boundary to the system. The key takeaway from this chapter includes the mathematical formulas that enable one to predict overall system performance and optimize designs.

Chapters 4−6 detail the refrigeration and power generation applications with analytical models to help quantify their performance. Chapter 4 focuses on refrigeration. Here, the design objectives are the coefficient of performance (COP) and the maximum cooling capacity Q_{max}. Unique characteristics of the relationship as the function of the thermoelectric element (leg) design parameters are shown. This chapter also shows the impact of several other features, such as performance limits with a large ZT, thin film devices, temperature dependence of material properties, and transient response. The performance of thermoelectric system is compared to vapor compression cycles. Finally, the integration of thermoelectric sub-cooler into a vapor compression cycle is discussed.

Similarly, Chapter 5 discusses maximum power output and material cost optimization in power generation. Exergy analysis is introduced to consider the quality of heat in the energy conversion process.

What-if analysis assuming a large ZT helps to understand the behavior of thermoelectric power generation as a thermodynamic system. The rest of the chapter discusses multi-segment modules, temperature dependent properties, and thermomechanical reliability for thermal cycling.

Chapter 6 explores applications for power generation. The thermoelectric topping cycle can enhance power plants, e.g. coal-fired steam turbines, by utilizing unused exergy. This concept can be applied to other types of power generation systems. This chapter also covers high-temperature waste heat recovery in industrial process, mid-temperature exhaust gas heat recovery from automotive vehicles, and solar energy harvesting.

Chapter 7 covers the emerging applications for thermoelectric energy harvesting. Power demands for wearable electronics and IoT devices are increasing significantly as batteries are not a long-term solution in these applications. Thermoelectric power generation from hot/cold water pipelines and from human body heat are discussed in detail.

Characterization methods are described in Chapter 8, which are essential for research and development of new materials and devices. First, characterization of individual material properties is covered. Subsequently, various advanced techniques based on micro- and nanodevices are shown. Harman method and impedance spectroscopy are also introduced since both are particularly useful to characterize thermoelectric devices.

Simulation tools are described in Chapter 9. The set of simulation tools are open for public through nanoHUB platform (https://nanohub.org/). Each software tool is written following the analytical models in this book. Interfacing these tools from devices (THERMO tool) to systems (ADVTEE and TEDEV tools) connected with the material property simulator (BTEsolver tool) are beneficial to understand the impact of design on efficiency, cost, and environmental impact. The graphical representation of the results helps to intuitively understand the key trade-offs. The nanoHUB cloud computing provides the background calculations for these Web-based tools.

Chapter 10 gives an outlook about the future of thermoelectrics in energy landscape based on the overall energy demand and the current status. A distributed power system is an ideal case where scalable thermoelectrics are fully utilized. The rest of this chapter provides authors' view on the most commonly asked questions: What are some of the promising directions to increase material's figure of merits, ZT? Will thermoelectrics become a major player in thermal energy conversion?

We hope readers will find our focused and selected topics useful along with detailed examples and online simulation tools. This book is dedicated to diverse researchers in energy conversion, thermoelectrics, micro/nanomaterials and devices, electronic and optoelectronic packaging, and thermal systems.

Finally, we wish to express our gratitude to our former colleagues and collaborators for their helpful discussions and collaborations in many parts of the book. Without their contributions, this book would never have been possible.

Kazuaki Yazawa
Je-Hyeong Bahk
Ali Shakouri

About the Authors

 **Kazuaki Yazawa** is a Research Professor at Purdue University (Indiana, USA). Prior to moving to academia, Dr. Yazawa spent about three decades at Sony Corporation (Tokyo, Japan) developing and implementing energy-efficient thermal management technologies for modern, computerized electronic devices. He earned his Ph.D. in Mechanical Engineering at Toyama Prefectural University while working at Sony. He has been active in the communities of thermal management and thermoelectric energy conversion not only in Japan and USA but also internationally. As an academic study, his research has focused on energy conversion from heat to electricity, exploring from fundamentals to system optimization.

 **Dr. Je-Hyeong Bahk** is currently an Assistant Professor jointly at the Department of Mechanical and Materials Engineering and the Department of Electrical Engineering and Computer Science, University of Cincinnati, OH, USA. He received his B.S. and M.S. from Seoul National University, South Korea, and his Ph.D. from the University of California, Santa Barbara in 2010, all in Electrical Engineering. Dr. Bahk has worked in the field of thermoelectrics for more than 10 years since his Ph.D. study, participating in numerous research projects on design and fabrication of efficient nanotechnology-based thermoelectric materials and devices. He has also developed several online simulation tools at nanoHUB.org that are capable of thermoelectric devices and materials simulation for research and education. His recent research interests include nanocarbon-based flexible thermoelectric materials and devices, various cooling technologies including CPU cooling and large-scale thermoelectric-based air-conditioning, and human body-heat energy harvesting for wearable electronics and sensors.

Ali Shakouri is the Mary Jo and Robert L. Kirk Director of the Birck Nanotechnology Center and a Professor of Electrical and Computer Engineering at Purdue University. He received his diplome d'Ingenieur in 1990 from Ecole Nationale Superiere des Telecommunications in Paris, France, and his Ph.D. in 1995 from California Institute of Technology in Pasadena, USA. He was a faculty at the University of California in Santa Cruz before moving to Purdue in 2011. His group studies nanoscale heat transport and electrothermal energy conversion to improve electronic and optoelectronic devices. They have also developed novel imaging techniques to obtain thermal maps with sub diffraction-limit spatial resolution and 800 ps time resolution. He is applying similar methods to enable real-time monitoring of functional film continuous manufacturing. He initiated a sustainable engineering curriculum and a California–Denmark summer program in renewable energies in collaboration with colleagues at UC Davis, Merced, DTU and Aalborg. More recently, he has focused on the scalable manufacturing of Internet of things devices and a regional testbed for smart manufacturing and digital agriculture in the 10 counties surrounding Purdue.

Acknowledgments

The authors acknowledge with thanks all financial support to conduct researches in thermoelectric materials, analytic modeling, device-level development, and system modeling and analyses.

The authors would also like to express special thanks to our collaborators in Thermionic Energy Conversion Center (MURI), Profs. John E. Bowers, Arun Majumdar, Timothy D. Sands, Arthur C. Gossard, Rajeev Ram, and Dr. Lon E. Bell.

The researches described in Chapters 4–6 were supported by the Center of Energy Efficient Materials, one of the Energy Frontier Research Centers (EFRC) of The Office of Science, U.S. Department of Energy. Also, the researches on thermoelectric heat pump (Chapter 4) and heat exchanger effectiveness (Chapter 3) were partially supported by the Center Excellence for Integrated Thermal Management for Aerospace Vehicles (CITMAV) as part of an industrial consortium project.

The authors acknowledge that the simulation tools (software codes) are outcomes of numerous collaborations with our former and current colleagues and collaborators. We are grateful for their contributions to the codes. Special thanks to Dr. Zhixi Bian, Ms. Megan Youngs, and Mr. Kevin Margatan for their major contributions.

Authors Kazuaki Yazawa and Ali Shakouri thank Drs. Kwon and Woolley (Saint Gobain) for their critical comments about the glass melt process and Dr. Caillat (NASA-JPL) for providing information on the high thermoelectric materials developed in NASA-JPL.

Author Kazuaki Yazawa thanks Cambridge Display Technology, Ltd., a company of Sumitomo Chemical group for the partial grant support on heat transfer modeling of wearable thermoelectric modules. Author KY also acknowledges the support from Nissan Chemical for their partial grant support for multi-layered thin-film thermoelectric power generators. In the project, fruitful discussion with Dr. Yee Rui Koh allowed us to gain some new knowledge.

Contents

Chapter 1

Introduction

1.1 Discovery of Thermoelectricity

The discovery of thermoelectricity was made by Thomas Johann Seebeck through his experiment in 1821 and later publication [1]. The experiment consisted of an electrical junction of two dissimilar compass magnets. Two different temperatures were applied across the junction, which created an electrical current through the materials as they form an electrical circuit. Seebeck initially thought it was a magnetic effect, but later corrected that this phenomenon was due to the voltage development in the system under temperature gradients. This effect was named after him as the *Seebeck effect* despite the fact that the beginning of the history of thermoelectricity was the report by Hans Christian Oersted in 1820, one year before Seebeck's report about the electromagnetism [2]. While Oersted investigated only on metal, Seebeck investigated enormous number of materials including semiconductors, which is the major class of material recent thermoelectric generators use. Earlier than Seebeck, Alessandro Volta also found the influence of temperature difference in a metal during his experiment on the electrical excitation of a frog leg back in 1794, but no physical explanation was provided at that time.

In a general definition, the Seebeck effect is 'the production of an electromotive force (EMF) and consequently an electric current in a loop of material consisting of at least two dissimilar conductors when two junctions are maintained at different temperatures' *Britannica* [3]. Interestingly, both of the early discoveries were made from an accidental observation of a current flow in a closed circuit rather than measuring a voltage. It was more obvious to measure or sense an electrical current flow rather than characterizing an open-circuit voltage at the time when electricity itself was not so familiar as it is in the present day.

Another important thermoelectric effect, the Peltier effect, was found by Jean Charles Athanase Peltier in 1834 [4]. Peltier observed that heat was generated at a junction of two dissimilar conductors when an electrical current was injected. He also clearly observed that either heating or cooling occurs at the junction depending on the direction of current flow.

Later, in 1851, William Thomson observed that heating or cooling occurs when an electrical current flows through a single conductor under a temperature gradient [5], which is now called the Thomson effect. The Thomson effect is an extension of the Seebeck and Peltier effects as they are inter-related. It is a generalization of the Peltier effect occurring within a single material when the Seebeck coefficient of the material changes with temperature.

The modern era of thermoelectricity is considered to during the 1950s–1960s after World War II. Russian Scientist Abram Fedorovich Ioffe contributed a significant work on thermoelectricity. Power generator based on the semiconductor thermoelectric materials [6] was developed and implemented with kerosene lamp and supplied power for radio receivers [7]. Cooling application was also actively investigated. These devices are the foundation of the modern design of thermoelectric modules. The concept of the material figure of merit (Z or non-dimensionalized ZT) of thermoelectricity was established by Ioffe and has been used to evaluate the materials on the same index. One disappointment in the era is missing the finding of the material with ZT well over 1.

Here, Z is the value defined by the following. Multiplied by the mean temperature at the operation, the ZT value interestingly becomes closer to near 1 for the material investigated at that time:

$$ZT = \frac{\sigma S^2}{\beta} T_{\text{mean}} \qquad (1.1)$$

where σ is electrical conductivity, S is Seebeck coefficient, and β is thermal conductivity. Recent research activities and achievement on thermoelectric materials are found in Chapter 9.

Based on the principles of thermoelectricity, this book is focused on how one can maximally translate the physical understanding into full industrial utilization. The key topics include the following:

(1) Thermoelectricity describes multiple physical phenomena related to the interactions and conversion between thermal and electrical energies. The thermoelectric effects are observed in semiconductors and conductors, and even in poor conductors, such as some oxide ceramic materials. The performance of the materials has not been improved much since the modern era of thermoelectrics in 1950s–1960s. Rather than focusing on material properties, we aim at maximal utilization of the contemporary thermoelectric materials for device and system performance.

(2) Due to the deep inter-relation between electrical and thermal transport in thermoelectricity, thermal design of the system is as critical as the electrical design. In applications like power generation or heat pump, co-optimization of electrical and thermal performances is essential to make a big difference in cost-effective performance measures, such as power output or cooling power per material volume or per cost.

(3) A thermoelectric energy conversion device is a solid-state system that relies on completely different physical principles compared to the conventional mechanical heat engines or vapor compression heat pumps. However, they equally obey both the first and second laws of thermodynamics in their internal processes. We view thermoelectric devices and systems as thermodynamic engines in this book.

(4) Owing to the simple structure of thermoelectric modules, extension to flexible or bendable design is possible for heat energy conversion from curved source surfaces. We discuss wide range of applications for on-demand and on-site energy harvesting that is suitable as a power source for the Internet of things (IoT).

1.2   Thermodynamic Cycles

The term 'heat engine' is associated with the concept of mechanical thermodynamic cycles, such as in gas turbines and automobile engines. The former typically uses the Brayton cycle [8] and the latter popularly uses the Otto cycle [9] or the Diesel cycle [10]. An ideal heat engine is called the Carnot engine, in which all processes in the cycle are reversible. It means that there is no energy loss both in mechanical and thermal forms. All these thermodynamic power generators share a process consisting of pressure and volume changes of the working fluid along with the temperature change due to the heat exchanges in the process. The change in pressure during the heat exchange processes in the working fluid results in useful mechanical work output, which drives an alternator to generate electric power. In the process of the thermal-to-mechanical energy conversion, the specific heat of the fluid takes an important role in creating a pressure change by temperature change in a control volume. Heat transfer occurring in the thermodynamic cycle is not necessarily directly connected to the mass transport induced by the pressure difference, hence the heat exchanger design is critical for the cycle. Figures 1.1 and 1.2 show the simplified and ideal Brayton and Otto cycles to emphasize their differences and similarities.

Fraction of the mechanical power is used to compress the working fluid and the rest of the mechanical power is utilized for motive power

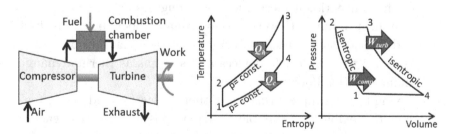

Fig. 1.1:   Brayton cycle for power generation, an example of continuous cycles by sharing a rotational shaft. Heat is incoming from the combustion chamber and released through the exhaust. In the cycle, mechanical work is applied at the compressor and recovered at the turbine by increasing gas power with heat input. Useful work is converted to electricity for a case of gas turbine generator.

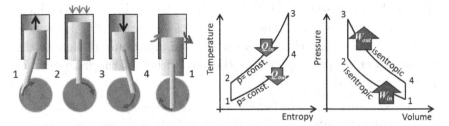

Fig. 1.2: Otto cycle for power generation, an example of a reciprocal piston gasoline engine. Heat is applied by igniting the fuel and air mixture. The flywheel at the bottom recovers the motion after the gas volume fully expands as the piston reaches the bottom end. The useful work is harvested as the rotation of the wheel.

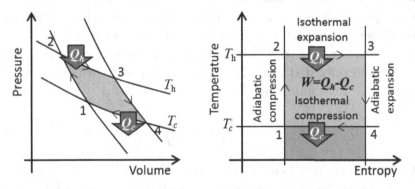

Fig. 1.3: Carnot cycle. The cycle consists of the adiabatic and isothermal heat exchange processes. The process has no heat loss so that it is perfectly reversible. The reverse flow process represents a refrigeration cycle.

to make a motion of mass, with options such as automotive vehicles or an alternator (electromagnetic motor) to generate electrical power, e.g. in a power plant.

Here, an ideal Carnot-cycle heat engine is briefly revisited (Fig. 1.3). In 1890, Nicolas Léonard Sadi Carnot stated the following proposition in his publication [11]:

The motive power of heat is independent of the agents employed to realize it; its quantity is fixed solely by the temperatures of the bodies between, which is effected, finally, the transfer of the caloric.

The motive power, in his era, specifically meant the power out of a steam engine, but the statement is generic enough to be applied to thermoelectric systems as well. Reversible heat engine follows an ideal thermodynamic process, where no heat loss occurs during the whole process. The entropy change by an external input only reflects to the useful power output that obeys the second law of thermodynamics as Carnot stated:

> *The efficiency of a quasi-static or reversible Carnot cycle depends only on the temperatures of the two heat reservoirs, and is the same, whatever the working substance. A Carnot engine operated in this way is the most efficient possible heat engine using those two temperatures.*

No thermodynamic cycles can generate useful work in efficiency larger than the Carnot limit. By denoting the hot reservoir temperature as T_h and the cold one as T_c, the relationship between the useful power output W and the heat input Q is described as the following for the reversible Carnot cycle at thermodynamic limit and the ratio η is called the Carnot efficiency or the Carnot limit, which is achieved only if the ZT is infinitely enhanced with the thermal contacts as reversible (thermal resistances are zero):

$$\eta = \frac{W}{Q} = 1 - \frac{T_c}{T_h} \tag{1.2}$$

An important idea here is that every thermoelectric power generator obeys the first (energy conservation) and second laws of thermodynamics.

1.3   Refrigeration and the Coefficient of Performance

Similar to power generation, thermodynamic refrigeration cycles are realized with compression/expansion and heat absorption/rejection. Vapor compression cycle is the most common process for refrigeration by taking the benefit of the large latent heat of vaporization. The cycle consists of at least one compressor. Ideal Carnot cycle has a pressure recovery mechanism, e.g. turbine on the opposite side of the cycle, which is also an ideal Brayton cycle (Fig. 1.4). A real vapor compression cycle typically has an expansion valve instead of a turbine in between the condenser and the evaporator heat exchangers (Fig. 1.5). The valve essentially does not recover the work applied by

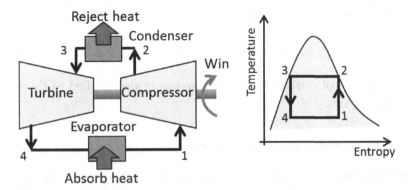

Fig. 1.4: Ideal Carnot vapor refrigeration cycle (reverse Brayton cycle). Hatched area in the temperature–entropy diagram shows the state of vapor–liquid mixture. The curve depends on the refrigerant used.

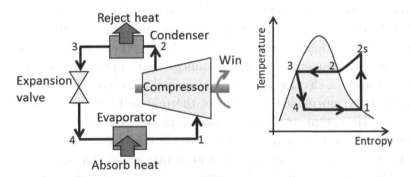

Fig. 1.5: Ideal vapor compression cycle with an expansion valve. The expansion valve allows temperature change by expansion but does not recover any internal energy. Compression through compressor is ideally isentropic, hence Point 2s on the right figure reflects the isentropic process. In this process, compression process from Point 1 to 2s is purely in a vapor state. Perfect condensation to liquid at the condenser is assumed.

the compressor, hence some of the cycle has unique additional mechanism to recover the useful energy. The refrigerant used as a working fluid for the cycle is typically chosen from one of the low global warming potential (GWP) fluids, e.g. hydrofluorocarbons (HFC) and hydrochlorofluorocarbons (HCFC).

The coefficient of performance (COP) is defined as the ratio of heat pumped $Q_{\text{pump}} = Q_{\text{rejected}} - Q_{\text{absorbed}}$ to the work used W_{in}, e.g. electrical power, such that

$$\text{COP} = \frac{Q_{\text{pump}}}{W_{\text{in}}} \qquad (1.3)$$

This measure of refrigeration performance is also applied to thermo-electric coolers.

1.4   Thermodynamics of Thermoelectrics

Although thermoelectric devices appear very different from the above mechanical thermodynamic cycles, their thermal behavior and functionality can be similarly described. The internal working fluid in thermodynamic cycles is replaced with electrical current in a thermoelectric device and the pressure changes correspond to voltage drops/gains. In this analogy, one big difference is that both thermal and electric transports occur in one single solid, called a thermoelectric leg, in a thermoelectric device, whereas they are separated in mechanical heat engines. In mechanical heat engines, a mechanical engine and an electromagnetic motor are connected by sharing the shaft. This 'morphological twin' of thermal and electrical transport leads to the compactness of thermoelectric devices. At the same time, however, it makes the interaction between the two transport regimes complex and makes the independent improvement difficult.

The key essential fundamentals of the uniqueness of thermo-electricity in physics are briefly described here. Thermodynamic point of view on thermoelectric energy conversion processes is also described in comparison to the mechanical thermodynamic processes mentioned earlier.

1.4.1   *Seebeck effect*

The Seebeck effect happens at the heated junction where the excess charged carriers are excited due to the doping to the semiconductor materials. The charged carriers over the intrinsic carrier density are electrons for n-type and holes for p-type. These two dissimilar materials have a contact called junction in a popular Π-shape structure of a thermoelectric module (Fig. 1.6).

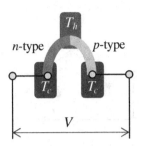

Fig. 1.6: Seebeck effect in one pair of n-type and p-type pair thermoelectric legs. The top side junction is heated to T_h and the open terminals are maintained at lower temperature T_c. A voltage potential V between the terminals is measured.

Due to the gradient of temperature, the generation of excess charge also shapes the gradient between T_h and T_c and results in the electrical potential V across the thermoelectric legs, which is considered as 1D. The voltage gain induced by the temperature difference is called Seebeck coefficient S [V/K]. Seebeck effect is considered for entirely across a 1D solid with a temperature gradient as given in the following equation. For the simplest case for solid bar, uniform cross-section, and isotropic material property with adiabatic boundary, the relation becomes linear. This process occurs much faster than the thermal response of the leg due to the large difference of inertia between the thermal and electrical capacitances:

$$V = \int_{T_c}^{T_h} (\Delta S) dT \tag{1.4}$$

1.4.2 *Peltier effect*

The Peltier effect is the reverse process referring to the Seebeck effect. At the junction of dissimilar materials, heating or cooling occurs induced by electrical current I going though in a circuit depending on the current direction (Fig. 1.7).

The product of temperature and Seebeck coefficient $\Pi = ST$ is equal to the Peltier coefficient:

$$Q = S(T_h - T_c)I = \Pi_{hc}I \tag{1.5}$$

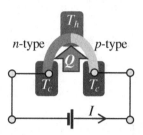

Fig. 1.7:   Peltier effect in one pair of n-type and p-type pair thermoelectric legs. The two terminals of the legs are connected so that the circuit allows electrical current I to flow. Depending on the flow direction, the heat flow is induced across the leg between the top junctions and the bottom terminals.

1.4.3   *Thomson effect*

The Thomson effect describes the heating or cooling of a current-carrying conductor with a temperature gradient. In this case, the Thomson effect is observed through the n-type and p-type legs with the existence of electrical current. The Thomson effect is an extension of the Seebeck and Peltier effects as they are inter-related, as stated earlier. It occurs in a single material when the Seebeck coefficient of the material changes with temperature. In a 1D system, total heat generation is found as follows:

$$Q = \rho \frac{d}{A} I^2 - \int \mu I dT \qquad (1.6)$$

where μ is Thomson coefficient related to Seebeck coefficient S as

$$\mu = T \frac{dS}{dT} \qquad (1.7)$$

For both power generation and refrigeration, the above effects are combined in the thermoelectric devices and induce the unique process. Here, thermodynamic cycle of thermoelectric power generator and chiller consisting of the above effects are shown in Figs. 1.8 and 1.9, respectively. These diagrams show the general outlines of the thermodynamic cycle and the details are discussed in Chapters 2 and 3, respectively [12].

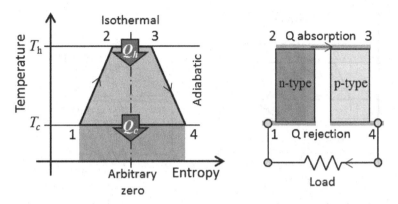

Fig. 1.8: Temperature–entropy diagram of generic thermoelectric power generator. Processes 1–2 and 3–4 correspond to n-type and p-type, respectively. In this view, only the hot side and cold side contacts of the thermoelectric are shown. With this ideal boundary constraint, the mid-point between two thermoelectric legs is considered as arbitrary zero in entropy for the analysis of single pair.

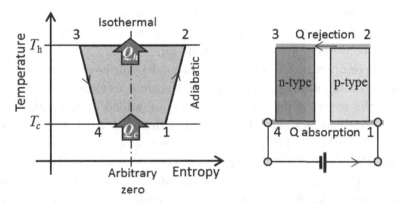

Fig. 1.9: Temperature–entropy diagram of generic thermoelectric refrigeration cycle at the maximum COP. Processes 1–2 and 3–4 correspond to n-type and p-type, respectively.

1.5 Role of Heat Transfer in Thermoelectric Systems

The diagrams shown in the previous section are conceptual as temperatures at both contacts of thermoelectric legs are given (fixed). In real systems, these temperatures are subsequently determined. The temperatures are altered not only by the external thermal contacts

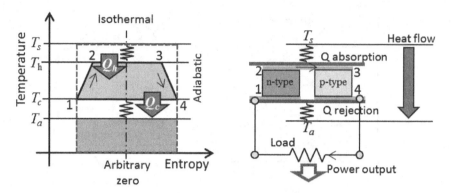

Fig. 1.10:   Temperature–entropy diagram of thermoelectric power generator systems. External and intrinsic heat transfer is critical for the design of system.

but also by the internal heat conduction as well as carrier heat transport. These heat transfer components must be taken into account in order to identify the equilibrium temperatures. A careful modeling of external heat transfer at both hot and cold sides is critical. In Chapter 3, heat transfer mechanisms related to the thermoelectric energy conversion are discussed in detail.

The temperature reservoirs at both sides can be assumed to have constant temperatures, T_s and T_a, all the time, as shown in Fig. 1.10. Here, the external thermal resistances are added between the heat source and the hot side of the TE legs denoted as ψ_h and between the cold side of the TE generator and the heat sink reservoir denoted as ψ_c. Previously shown diagrams of thermodynamic cycles and systems in Figs. 1.8 and 1.9 are then modified to Fig. 1.10. The power generated in the closed loop 1–2–3–4 is modified due to the external thermal resistances and the system efficiency is reduced according to the larger denominator in Eq. (1.2) in comparison to the perfect contact, zero thermal resistances.

In a practical system consisting of finite thermal resistances ($\psi_h > 0$, $\psi_c > 0$), heat flow is limited by the external temperature gradients in both hot and cold sides:

$$Q_h = \frac{(T_s - T_h)}{\psi_h}, \quad Q_c = \frac{(T_c - T_a)}{\psi_c} \tag{1.8}$$

As Eq. (1.4) suggests, the power output generated by the thermo-electric would be parabolic to the temperature gradient across the leg; $(T_h - T_c)$:

$$w \propto \frac{V^2}{R} \propto (T_h - T_c)^2 \qquad (1.9)$$

There exist two extreme cases: (1) One case is when $(T_h - T_c)$ goes to zero, then the power output obviously becomes zero by Eq. (1.8); (2) another extreme case is when $(T_h \to T_s)$ *and* $(T_c \to T_a)$, where the external resistances become zero and $(T_h - T_c)$ goes to its maximum. In the second case, power output appears to monotonically increase as shown in Eq. (1.9), but at the same time, heat flow diminishes according to Eq. (1.8). Therefore, the temperature set at the maximum power output should exists between the above two extremes. This also means that the thermal resistances must be properly designed for maximizing power or minimizing the power cost (\$/W). The optimization details will be discussed later in Chapter 5. The key takeaway here is that the power output of the system is strongly and directly related to the external and intrinsic heat transfer of thermoelectric modules.

In the refrigeration cycle, the direction of the pumped heat flow is in reverse to that of the passive heat conduction, which thus requires a work to do the active heat pumping, i.e. the electronic power externally supplied. Let the temperature of heat source be T_a, which is typically lower than that of the other reservoir (heat sink) to pump the heat to. Let the temperature of the heat sink be T_s. Obviously, temperature differences from T_a to T_c and from T_h to T_s must be positive for the heat conduction:

$$T_a - T_c > 0, \quad \text{and} \quad T_h - T_s > 0 \qquad (1.10)$$

Figure 1.11 shows the temperature profile and entropy relation in this system with finite external thermal resistances. The thermoelectric legs must internally hold the temperature gradient larger than the target temperature difference created between the reservoirs, i.e. $> T_s - T_a$. By operating the thermoelectric, the system

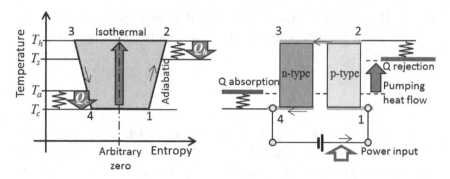

Fig. 1.11:   Temperature–entropy diagram of thermoelectric refrigeration system at the maximum COP. Temperature gradient internally across the leg must be larger than the externally given temperatures. External and internal heat transfer is critical for the design of the refrigeration system.

COP in Eq. (1.2) becomes smaller since the numerator of the COP must be smaller than the case of fixed terminal temperatures. In this case, $Q_{\text{pump}} = Q_c$.

Heat fluxes, Q_c, Q_h shown in Fig. 1.11, are the critical drivers of heat flow through the heat pump. If Q_c is small, the capacity of refrigeration is small. If Q_c needs to be too large, the power consumption goes extremely large, hence the Joule heating (quadratically proportional to the power consumption) limits the Q_c itself. Therefore, there must be an optimum design in between the two extreme cases. Typically, there are two design criteria, one for maximum COP and the other for maximum cooling power or cooling capacity for a given temperature target.

1.6   Scalability of Thermoelectrics

A thermoelectric generator performs the same function as a heat engine in the power generation mode and the same function as a vapor compression cycle in the heat pump or cooling mode. Prior to looking into the technical details, a bird's eye view is briefly discussed here regarding the scale of performance in potential thermoelectric applications.

Following are some of the several advantages of thermoelectric devices compared to conventional mechanical heat engines, although for the most cases, their relatively moderate energy conversion efficiency is a weakness:

(1) noiseless essentially no moving parts required [13];
(2) no oil/grease and less maintenance required [14];
(3) a variety of material options for broad range of temperature [15];
(4) flexible (polymers, fibers) for compliance to curved surfaces [16, 17];
(5) low profile — typically thickness less than 5 mm;
(6) highly scalable with arrays of smaller devices to cover a larger area.

The beauty of thermoelectric system is its scalability. Modular design for the arrays, e.g. if a 1 cm × 1 cm unit generates 1 W, could generate 10 kW with 1 m^2 area of arrays, as far as the hot and cold working fluid temperatures are uniform and the blower and pump are externally available. Note that the unit power is determined by the hot and cold reservoir temperatures and unit power varies per applications.

Figure 1.12 shows the power density spectrum for a broad scale of thermoelectric applications. Thermoelectric generator built in a wristwatch [18, 19] can harvest the body heat from the wrist skin. Based on the heat dissipation rate (40–600 W/m^2), generating power is estimated as a less than 1 μW/cm^2 based on the temperature difference of 0.4°C across the thermoelement. For higher temperature applications, a topping cycle thermoelectric generator [20] for steam turbines in a large-scale ($\sim$0.5 GW/m^2 class) coal-fired power plant was analyzed. Around 1400°C of hot gas temperature is available in the boiler chamber, while only 550–600°C temperature is the maximum of steam for maintaining a long-term mechanical reliability of the turbine blades. As a result, the ideal thermoelectric design ends up with six figures different in power output per unit area compared with body heat harvesting power generation, while both designs are based on the exact same principle and model.

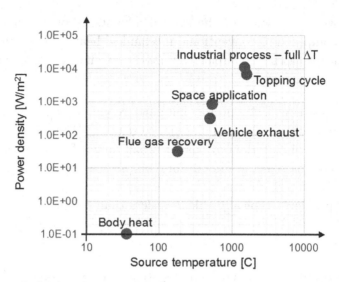

Fig. 1.12:   Scalability diagram. Calculated harvesting electrical power per unit applicable area from thermoelectric systems as a function of heat source temperature against the room temperature 20°C at the other side. The scale of power density ranges six figures, while commercial solar PVs are generated in quite a narrow range of 100 W to few hundreds of Watts per area (solar irradiation coverage).

1.7   Societal and Environmental Impact

Conventional power generation systems or further energy-efficient combined heat and power (CHP) systems produce useful electricity and heat energy in technically reasonable thermodynamic efficiencies. However, in the global view of preserving and protecting environment and even from the energy economy point of view, it must be evaluated far behind. Firstly, majority of the power supply is based on the combustion of fossil fuels, while renewable technology does not quickly penetrate. Apart from the worries about the exhaustion of natural resources, limiting accumulated CO_2 emission (2/3 of contribution to greenhouse effects) is also a global issue. The regulations for future near-term and long-term are quite strict for the use of such biased energy resource dependence. Under the condition, thermoelectric may assume an important role in improving the total system energy efficiency. Despite the moderate energy conversion efficiencies, the thermoelectric power generators can be utilized virtually anywhere as far as heat is just wasted. Some analyses have already been performed and are available in the literature. Yazawa *et al.* investigated a thermoelectric topping cycle for a Rankine cycle discussed in the previous section and a modified Brayton cycle [21] or industrial heat recovery [22]. Chen *et al.* reported the impact of utilizing thermoelectric for 200°C temperature range with off-the-shelf modules for harvesting waste heat and heat recovery in a power generation cycle [23]. de Castro Villela *et al.* investigated ecological impact [24] and Patyk investigated environmental and economic perspective [25]. The efficiency improved by introducing thermoelectric has been reported at somewhere between four and nine points in addition to the current power generation or CHP systems.

Thermoelectric could also be used as solar energy harvesting, especially using concentration. As photovoltaic cells are highly temperature-dependent, cooling the back side is a meaningful approach. Thermoelectric is a moderately thermally conducting device that can be used simultaneously as a passive heat exchanger and an active extra-power generator. Since there is a trade-off relation between the PV and thermoelectric generations, there is an optimum design combination with PV cells [26]. By utilizing a selective optical absorber and reflector, nearly 40% of the solar energy can be harvested as a heat input, as the remaining 60% is harvested by a photovoltaic [27]. Thermo-photovoltaic is another approach, for which literature is available [28].

References

[1] T. J. Seebeck, Magnetische polarisation der metalle und erze durch temperatur-differenz, in *Ostwald's Klassiker der Exakten Wissenshaften*, Nr. **70**, 1822–1823 (1895).

[2] Abram F. Ioffe, The revival of thermoelectricity, *Sci. Am.*, **199**(5), 31–37 (1958).

[3] The editors of Encyclopaedia Britannica, 'Seebeck effect', https://www.britannica.com/science/Seebeck-effect.

[4] J. C. A. Peltier, Nouvelles expériences sur la caloricité des courants électrique, (New experiments on the heat effects of electric currents), *Annales de Chimie et de Physique*, **56**, 371–386 (1834).

[5] W. Thomson, On a mechanical theory of thermo-electric currents, *Proc. R. Soc. Edinb*, **3**(42), 91–98 (1851).

[6] R. R. Heikes and R. W. Ure Jr., *Thermoelectricity: Science and Engineering*, Wiley Interscience, New York, 405 (1961).

[7] M. V. Vedernikov and E. K. Iordanishvili, AF Ioffe and origin of modern semiconductor thermoelectric energy conversion, *Proc. Seventeenth International Conference on Thermoelectrics* ICT98, 37–42 (1998).

[8] M. J. Moran and H. N. Shapiro, Gas power systems, 9.6 air-standard Brayton cycle, in *Fundamentals of Engineering Thermodynamics*, 2nd edition, John Wiley & Sons, 374–387 (1988).

[9] M. J. Moran and H. N. Shapiro, Gas power systems, 9.2 air-standard Otto cycle, in *Fundamentals of Engineering Thermodynamics*, 2nd edition, John Wiley & Sons, 360–364 (1988).

[10] M. J. Moran and H. N. Shapiro, Gas power systems, 9.3 air-standard diesel cycle, in *Fundamentals of Engineering Thermodynamics*, 2nd edition, John Wiley & Sons, 365–368 (1988).

[11] S. Carnot, *Reflection of the Motive Power of Heat: From the Original French of N.-L.-S. Carnot*. John Wiley (1897).

[12] H. T. Chua, K. C. Ng, X. C. Xuan, C. Yap and J. M. Gordon, Temperature-entropy formulation of thermoelectric thermodynamic cycles, *Phys. Rev. E*, **65**(5), 056111 (2002).

[13] G. Min and D. M. Rowe, Experimental evaluation of prototype thermoelectric domestic-refrigerators, *Appl. Energy*, **83**(2), 133–152 (2006).

[14] J. G. Stockholm, Reliability of thermoelectric cooling systems, in *Proc. of the 10th Int. Conf. on Thermoelectric Energy Conversion*, 27–31 (1991).

[15] M. S. Dresselhaus, G. Chen, M. Y. Tang, R. G. Yang, H. Lee, D. Z. Wang, Z. F. Ren, J-P. Fleurial and P. Gogna, New directions for low-dimensional thermoelectric materials, *Adv. Mater.*, **19**(8), 1043–1053 (2007).

[16] B. Zhang, J. Sun, H. E. Katz, F. Fang and R. L. Opila, Promising thermoelectric properties of commercial PEDOT: PSS materials and their Bi2Te3 powder composites, *ACS Appl. Mater. Inter.*, **2**(11), 3170–3178 (2010).

[17] K. Yazawa and A. Shakouri, Heat transfer modeling for bio-heat recovery, In *Thermal and Thermomechanical Phenomena in Electronic Systems (ITherm)*, 2016 15th IEEE Intersociety Conference, IEEE, 1482–1488 (2016).

[18] M. Kawata and A. Takakura, Thermoelectrically powered wrist watch, U.S. Patent 5889735, issued March 30 (1999).

[19] S. Watanabe, Wrist watch having thermoelectric generator, U.S. Patent 6304520, issued October 16 (2001).

[20] K. Yazawa and A. Shakouri, Thermoelectric topping cycles with scalable design and temperature dependent material properties, *Scripta Mater.*, **111**, 58–63 (2016).

[21] K. Yazawa, T. S. Fisher, E. A. Groll and A. Shakouri, High exergetic modified Brayton cycle with thermoelectric energy conversion, *Appl. Therm. Eng.*, **114**, 1366–1371 (2017).

[22] K. Yazawa, A. Shakouri and T. J. Hendricks, Thermoelectric heat recovery from glass melt processes, *Energy*, **118**, 1035–1043 (2017).

[23] M. Chen, H. Lund, L. A. Rosendahl and T. J. Condra, Energy efficiency analysis and impact evaluation of the application of thermoelectric power cycle to today's CHP systems, *Appl. Energ.*, **87**(4), 1231–1238 (2010).

[24] de Castro Villela, Iraídes Aparecida and José Luz Silveira, Ecological efficiency in thermoelectric power plants, *Appl. Therm. Eng.*, **27**(5–6), 840–847 (2007).

[25] A. Patyk, Thermoelectric generators for efficiency improvement of power generation by motor generators–environmental and economic perspectives, *Appl. Energ.*, **102**, 1448–1457 (2013).

[26] K. Yazawa, V. K. Wong and A. Shakouri, Thermal challenges on solar concentrated thermoelectric CHP systems, in *Proc. of the 13th IEEE Intersociety Conference on Thermal and Thermomechanical Phenomena in Electronic Systems*, pp. 1144–1150 (2012).

[27] P. Bermel, K. Yazawa, J. L. Gray, X. Xu and A. Shakouri, Hybrid strategies and technologies for full spectrum solar conversion, *Energ. Environ. Sci.*, **9**(9), 2776–2788 (2016).

[28] D. Narducci, P. Bermel, B. Lorenzi, N. Wang and K. Yazawa, Hybrid photovoltaic–thermoelectric generators: Theory of operation, in *Hybrid and Fully Thermoelectric Solar Harvesting*, Springer, Cham, 91–102 (2018).

Chapter 2

Thermal Energy Conversion in Photonics and Electronics

This chapter describes the thermoelectric applications in photonics and electronics, including energy-friendly cooling, device thermal management, on-demand power generation, and waste heat recovery. The examples show how thermoelectric works to accomplish individual objectives including several technology readiness levels from concept to actual implementation. The most common market of thermoelectric technology commercialization is the cooling of electronics and photonics devices. The advantages of thermoelectrics are compactness, lightweight, quick response, precise temperature controllability, and scalability for the thermal solution.

2.1 Energy-Friendly Cooling of Systems

Due to the monotonic increasing of heat dissipation per unit area or unit volume [1], requirements of thermal management in photonics and electronics went far beyond the capacity of passive air cooling, e.g. natural convection. Active cooling with forced air or water convection has already been taking a significant role, especially in computers and digital equipment for the past few decades. The hierarchy of thermal management from the device, board, equipment cabinet, and the system for data centers or information technology (IT) centers is becoming more energy-friendly through significant technological development [2]. However, power consumption of active cooling air conditioning for megawatt devices such as computers pose a large-scale challenge. The cooling system requires not only power cost but also a large footprint for the equipment and the space for maintenance. Utility power such as the cooling power is a part of the system power usage effectiveness (PUE) [3] of the facility in data centers and the utility power is not small. Recently, however, major data center providers are developing technology to reduce the thermal management power, even to the point of investigating the possibility of net-zero energy data centers [4]. Active cooling for data centers consists of chiller units and circulating working fluid, either air or water, through the computing enclosures. The chillers are typically vapor compression cycles (VCs) that are centralized and working efficient (Fig. 2.1). However, the working fluid travels a long distance so that the heat loss along the transmission of chilled working fluid is non-negligible.

Thermoelectric cooling is theoretically replaceable in the VC, while realistic cooling heat $Q[\text{W}]$ per power consumption $W[\text{W}]$, called coefficient-of-performance (COP), is moderate and lower than VCs:

$$\text{COP} = Q/W \tag{2.1}$$

The advantage of using thermoelectric cooling can be demonstrated on a smaller scale and hence adaptable to a distributed system [5]. All cooling units can be brought into individual racks or cabinets so that heat loss and fluid dynamic loss along the working fluid path in ducts or pipes are eliminated. A comparison of COP with VCs is discussed in detail in Chapter 4.

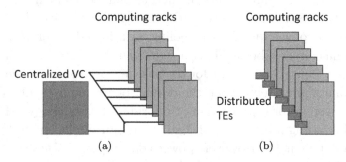

Fig. 2.1: Data center computing racks cooling with (a) centralized VC and (b) distributed cooling with thermoelectrics (TEs).

Following is an example of an energy-efficient cooling of a system by utilizing the thermoelectric hotspot cooling. The combination of general air conditioning for the racks and a focused liquid cooling for the high heat flux devices on board may yield energy-efficient cooling [6]. Microchannel cooling of CPU/high-performance chip has been considered as an effective cooling method targeting the hotspot, while the thermal management of other electronics on the board, such as blade server, is performed by convective conditioned airflow by cooling fans. Sensible cooling capacity with microchannel heat sink is quite high and proportional to the flow rate. However, the convection heat removal from the channel wall to water is limited. For example, Nusselt number is constant for fully developed laminar channel flow [7] and only proportional to the (4/5)th power of flowrate for turbulent flow, e.g. Sieder and Tate correlation [8], where the required pump power increases quadratic to the flow rate. This means that COP rapidly decreases at a certain point of thermal management, while such electronic devices are required to maintain a temperature not more than the allowable maximum.

Using a thermoelectric size of a hotspot in conjunction with water-cooled microchannel heat sink could save significant electrical power for pumping water [9]. The study case has a 0.5 mm × 0.5 mm hotspot with heat flux of 600 W/cm^2 additionally on a 100 mm^2 chip uniformly heated at 100 W. The thermal management requirement is to maintain the hotspot and chip temperature at 85°C or less. The comparison between cases with and without thermoelectric hotspot cooling shows significant difference in power consumption in system cooling, which is the inversely proportional to COP. Figure 2.2 shows the COP depending on the design conditions (thermoelectric leg thickness d[m] and electrical current I[A]). The utilization of thermoelectric right on to the hotspot will achieve 10 times the efficient cooling.

Laptops and handheld/portable electronics, even tablet or smart phones, tend to increase the peak operation power, especially on game operations or video/audio streaming, where heat is generated beyond the passive cooling limit of air convection. Hence, one or more air movers are potentially needed. Saving power consumption for the

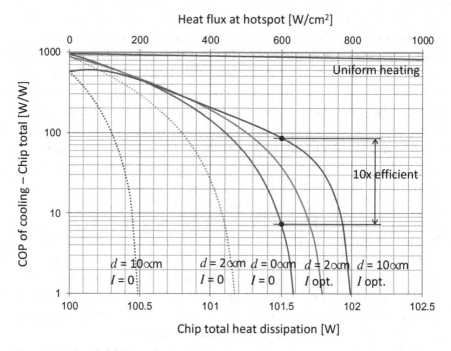

Fig. 2.2: Total COP of chip cooling with thermoelectric hotspot cooling. The conditions $d = 0$ and $I = 0$ imply that there is microchannel alone and no thermoelectric hotspot cooler is attached. All thermoelectric cases $(d > 0)$ show lower COP in small heat flux on hotspot but take over the baseline $(d = 0)$ at around 250 W/cm^2 and beyond with optimum drive current (I_{opt}).

active thermal management becomes more important for battery saving and thin and lightweight solution for these portable electronics.

In general, the popular method for cooling electronic enclosures is to install an air mover (cooling fan) with a heat sink, where power consumption to drive the fan is unavoidable. This mechanism is an energy conversion process along with thermodynamic behavior and it works effectively for the thermal management. The COP for this case is commonly shared with the same definition as a refrigeration cycle. The value of COP is found by calculating heat removal $Q[\text{W}]$ over motive work of air $W_{\text{air}}[\text{W}]$ and power input used for the heat removal $W_{\text{refrige}}[\text{W}]$. The W_{air} is needed to take over the resistance of convective fluid flow; $W_{\text{air}} = \Delta P \times \dot{V}_{\text{air}}$, where $\Delta P[\text{Pa}]$ is pressure drop across the flow passage and $\dot{V}_{\text{air}}$ [m^3/s] is volumetric flow rate of air. The work is done by a driving fan with motor efficiency η of transforming electrical power to the kinetic power. The W_{refrige} is the electrical power to drive the thermoelectric, vapor compression cycle, or similar cases:

$$\text{COP} = \frac{Q_{\text{removal}}}{(W_{\text{air}}/\eta + W_{\text{refrige}})} \tag{2.2}$$

This parameter seamlessly provides a fair comparison of technologies regardless of the refrigeration cycle or air/fluid convection cooling. In this case, the COP value ranges from 10 to sometimes over 100.

2.2   Device Cooling

Space cooling of a data center or air cooling of computer racks does not necessarily mean that the temperatures of critical hotspots are properly managed. Focused thermal management in addition to such air conditioning may be required. In addition, the heat flux of such a device still keeps evolving by time, following Moore's law [10]. Advanced finer technology is always desired from a cost perspective and at the same time pushing power density higher. Thermoelectrics cooling for an ultra-high heat flux is not a suitable application, especially over 1 kW/cm^2 despite the expectation of spot cooling. This is due to the fundamental heat flow in reverse direction against the heat removal by intrinsic thermal conduction. Typical thermal conductivity of thermoelectric material ranges 2–5 W/(m K), where the inverse thermal gradient is created by electrical current. As a result, a large amount of current is required for pumping such high heat flux, where a large amount of Joule heating happens. Thermoelectric cooling of hotspots with heat spreading throughout the semiconductor substrate by three-dimensional thermal diffusion has been studied [11] based on CMOS technology.

More recently, mobile communication infrastructure consumes more and more power as 5G wireless high-speed signal protocols began to take place. Thermal management is a matter of concern not only for the stations and commutation boxes but also for high-frequency power transistors made of wide band gap materials. In such devices, the thermal problem raises the scaling challenge of cooling in

sub-micron and sub-microsecond pulse heating. The transistor structure of high-power and high-speed switching devices, such as gallium nitride (GaN) or silicon carbide (SiC) in high-electron mobility
transistors (HEMTs), is designed with 2D electron gas (EG), which
occurs with a field bias underneath an insulating barrier layer of aluminum gallium nitride (AlGaN). 2D EG occurs within the thickness
of sub-microns and length of a few hundreds of microns. Gate lines
for field biasing are typically integrated as comb configuration into a
monolithic microwave integrated circuit (MMIC) chip (Fig. 2.3).

As with silicon-based thermoelectric for monolithic integration
of thermoelectric coolers [12], thermoelectric performance has been
investigated for GaN and AlGaN materials [13]. In this small scale,
thermal conduction is a critical component. High thermal conductivity material is the solution to bring out the heat as much as
possible. Hence, a thermoelectric device embedded underneath the
transistors [14] may be one of the potential solutions for hotspot cooling. Near-junction cooling is a harder challenge, where heat diffusion
no longer effectively works due to the impact of sub-continuum to
quantum scale on apparent thermal conductivity. In a research program conducted by the Defense Advanced Research Projects Agency
(DARPA), a microfluidic cooling with thermoelectric spot cooling

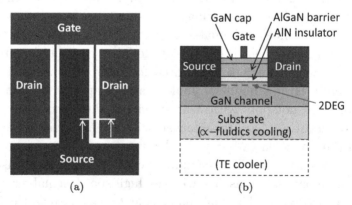

Fig. 2.3: Schematic diagrams of (a) GaN HEMT device top view and (b) the heterostructure of the transistor for 2D electron gas (EG) cross-section and zoom-up.
Microfluids cooling could be embedded in the substrate layer. Package-level TE
liquid cooler integration may work in conjunction with microfluidics.

was investigated [15]. The pre-cooling of fluid is a potential promising combination for providing benefits derived from both microfluidics and thermoelectric cooling.

2.2.1 *Cooling of photonics*

Coherent light beams generated from a laser diode (Fig. 2.4) is significantly valuable to transmit and receive high-speed signals with less noise compared to electron transport through electrically conductive wires. In high data rate signal transmission, as with a laser device, self-heating as a dissipating power becomes a technical challenge even when electrical to optical energy conversion is likely over 50% (depending on the capacity and operation condition).

Laser projectors for large screen displays are one of the consumer or commercial electronics where the luminescence from laser diodes dominantly define the range of performance. Power consumption of the semiconductor laser diode is relatively lower than halogen lamps and the operation temperature is also lower. Still the need for a cooling mechanism is not eliminated. One crucial difference that does

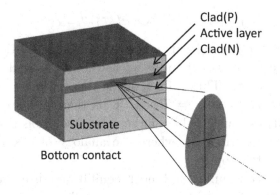

Fig. 2.4: Sample schematic diagram of edge emitting laser diode. The active layer placed between p-type and n-type material to make the band gap to generate photons and the lateral plane structure helps to resonate the photons. Finally, the coherent photons are emitted from the edge of the device. In most of the cases, thermal management device is attached to the bottom contact.

Examples are found everywhere, such as Ref. [16].

exist is that the cooling device can be much closer to the core of the heat generation in laser diodes, hence the thermoelectric spot cooling thermal management is more suitable.

A high-power laser exhibits transient heat generation and the thermally induced wavelength shift turns into a big challenge. In response to the big power pulse input, the temperature of junction quickly changes where the wavelength is very sensitive to temperature. Therefore, the thermal management must be responsive to a high-frequency operation. There has been some study of laser diode cooling using microchannels [17, 18]. However, there are a couple of fundamental limitations with microchannels since (1) removing heat flux is limited by the temperature gradient and (2) transient response of thermal diffusion is orders of magnitude slower than electrothermal generation. When a step heat is applied, the temperature responds in exponential manner, which is determined by the thermal resistance through the heat flow path and the thermal capacitance of the laser diode component. This temperature response is expressed in Eq. (2.3), which is a solution of 1D heat diffusion equation in a uniform and isotropic thermally conductive material. The diffusion equation is found in a regular heat transfer book:

$$\theta(t) = (1 - \exp(-t/\tau)) \, \theta_{t \to \infty} \qquad (2.3)$$

where $\theta(t)$ is the normalized temperature change by the step heat input that begins at $t = 0$ and $\theta_{t \to \infty}$ is the normalized equilibrium temperature change. This is for a single thermal mass system with time constant τ, where $\tau = R \times C$. C is thermal capacitance and R is thermal resistance. This mathematical solution is determined by Laplace transformation, where the analogy of electrical circuit conveniently often applies.

Instead of continuum and near equilibrium heat removal, by mass transport of a single-phase working fluid, the flash cooling approach was investigated to make the temperature of the heat source stable [19]. Synchronizing with the step heat input, the temperature of working fluid immediately decreases by switching to low pressure. Evaporation occurs instantaneously and removes the large amount of heat according to the heat of vaporization of the working fluid.

The uniqueness of thermoelectric cooler is found in time response to the step heat input. The Seebeck effect happens immediately as quick as electrical step input, whereas temperature change by internal Joule heating responds with the involving thermal mass (RC). A direct cooling of laser diode with thermoelectric was investigated in Ref. [20], the response to the pule input works well. But the accumulating heat residue by internal Joule heating in repeating pulse input is a critical issue that remains. The benefit of the combination of thermoelectric cooling and the phase change heat removal is the precise and high-frequency control of temperature because of narrower temperature range of heat removal section and the quick electrical rather than thermal response.

Another approach of cooling laser device has been investigated by using a quantum well structure [21], which moderately improves the allowable current density due to the leakage current threshold determined by the material property.

Light emitting diode (LED) display/TV is dominating the market of home entertainment displays. Most 'LED displays' have back-lit LEDs and then the color and brightness of individual pixels are controlled by a liquid crystal display (LCD) panel. Thermal management of backlight device is important, where thermal profile needs to be monitored for the uniform brightness of a large screen [22]. Thermoelectric application has not been seen in the literature for backlight cooling. The reason could potentially be the cost of thermal management. However, dynamic temperature control for the backlight might be utilized as a thermoelectric solution.

2.3 On-demand/On-site Power Generation

This section deals with the power generation for the device which requires the power supply on site and on demand. There are two major examples shown here.

2.3.1 *RTG in space satellites*

Multi-mission Radioisotope Thermoelectric Generator (MMRTG) was developed to provide a reliable power source for a long time

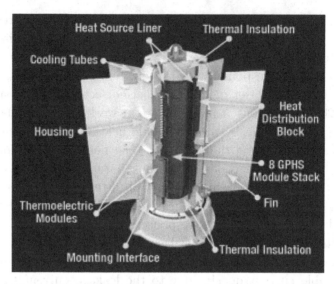

Fig. 2.5: Multi-mission Radioisotope Thermoelectric Generator (MMRTG) for space satellite [24]. Heat energy is extracted from Plutonium-238 and converted to 110 W of electricity with thermoelectric generators.

period as built-in power for space satellites by the Jet Propulsion Laboratory (JPL) of the National Aeronautics and Space Administration (NASA) (Fig. 2.5). Some other RTGs were developed in the institutions of other countries. The efficiency of this energy conversion system is estimated at 5.5%. The thermoelectric legs consist of the lead telluride (PbTe) n-type legs and the segmented lead antimony telluride (PbSnTe) with $(AgSbTe_2)_{0.15}(GeTe)_{0.85}$ called TAGS for p-type legs [23]. Plutonium ($Pu^{238}O_2$) is used as the source of heat, in which power density is 0.54 W/g with the half-life 87.7 years. The temperature at the legs of the thermoelectric depends on the thermal resistance of the thermal path. In outer space, final heat dissipation is only provided by the heat radiation to the temperature in outer space (2.73 K) and no convection occurs. Hence, the big fin radiators are necessary.

2.3.2 *Radioisotope pacemaker*

In the 1970s, pacemakers powered by nuclear (radioisotope) energy with a thermoelectric generator were developed and used as medical

Fig. 2.6:   Plutonium Powered Pacemaker with thermoelectric (1974).
Source: Courtesy of Oak Ridge Associated Universities.

devices (Fig. 2.6). Despite the concern of the potential risk of being exposed to radio toxicity, the device helped patients to extend their life expectancy. In less than a decade, the power source was replaced by lithium batteries and nuclear power has no longer been used for this purpose since then.

Due to the risks of enclosing nuclear isotope materials in a capsule, the device has fallen out of use in the commercial domains. However, thermoelectric power generation from body heat is still under investigation in scientific research [25].

More recently, research on harvesting body heat recovery for powering pacemakers has been explored [26]. This approach became realistic once the state-of-the-art electronics succeeded to dramatically reduce the power consumption of the pacemaker.

In harvesting body heat for powering electronics, a couple of thermoelectric-driven wrist watches were developed in the late 1990s [27, 28]. At the time of their initial development, the mechanical movement was driven by a motor with a thermally generated power from a temperature difference between the exterior metal casing and the wrist skin surface. For technical details, see Chapter 6.

2.4   Scavenging Waste Heat

In relatively large systems like a data center or a high-performance computing facility, large amount of heat is unintentionally generated at temperature ranging from 40°C to 75°C. The heat amount is possibly hundreds of megawatts for one site since the IT power

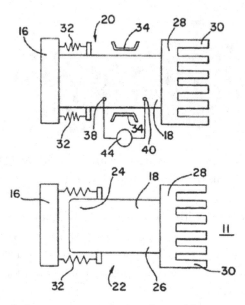

Fig. 2.7:   Low-grade heat recovery using mechanical hysteresis with shape memory alloy. See details in US Patent 6,668,550.

likely achieves a range of gigawatts. Despite the scale of energy, these are low grade and typically unused other than as a designed part of a combined heat and power (CHP) system. Low-grade waste heat recovery/scavenging is valuable as far as the cost for the equipment is paid off by saving of grid power (Fig. 2.7). For the low-grade heat recovery, organic Rankine cycles (ORCs) and absorption chillers (ACs) have been actively investigated [29, 30]. Organic Rankine cycle is essentially a low-temperature version of steam turbines (see Section 1.3). The refrigerants are typically organic macromolecules so that the vaporization temperature is sufficiently low to meet the temperature of the waste heat. The disadvantage of ORC is the smaller sensible and latent heat and the lower thermal conductivity than the working fluids that are proper for higher temperature range, such as steam. An absorption chiller is driven by heat to pump another heat from an evaporator into lower temperature. Hence, the process is using waste heat to make chilled air or water for cooling. The advantage is the complexity of the cycle so that a certain level of scale is required ($\sim$ 100 kW or larger).

A few other approaches such as a thermomechanical expansion cycle using shape memory alloy [31] has been investigated, but no commercial development has not been seen so far. Vanadium oxide (VO_2) was revealed to have temperature hysteresis in electrical and optical properties, which could also formulate a cycle to generate useful energy from waste heat [32].

In a smaller scale, waste heat from semiconductor chips can be recovered by using thermoelectric to feed an additional power as auxiliary back into the system. The unique requirement in this challenge is the physical size. Scavenging mechanism must be in the same size as the heat source. Waste heat from a CMOS CPU chip in a laptop computer was recovered as electrical power to drive a cooling fan [33]. Overall heat dissipated from the chip was converted to electricity and used to manage the temperature with no external power. A thermoelectric module was placed between the heater (imitating CPU) and a heat sink with a cooling fan (Fig. 2.8). With increasing power consumption of CPU up to 25 W, temperature gradient across

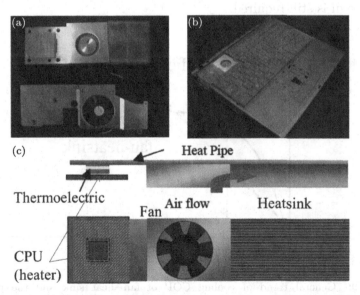

Fig. 2.8: Picture and diagram of thermoelectric fan–heat sink cooler. (a) Experimental sample (upper) and original heat sink (lower), (b) built in a laptop body, and (c) diagram of the structure.

the attached thermoelectric increases. The cooling fan is electrically driven by the thermoelectric. Then, the cooling fan drives airflow sufficiently to maintain the temperature of CPU as it generates heat. The cooling capacity increases as the flow rate of air increases. In this system, the higher the heat generation by CPU, the larger the flow rate, and hence the larger the heat removal. At an equilibrium, the heater maintained a temperature at a given level.

In general, adding the thermoelectric device passively increases the total thermal resistance in the heat removal path. In this case, however, the scavenged power drives the fan stronger to reduce the heat sink thermal resistance dynamically. The design point may exist, where the COP of fan and heat sink matches the inverse of energy conversion efficiency at the allowable maximum temperature $T_j(\Delta T = T_j - T_a)$. Figure 2.9 shows the general trend of the fan−heat sink COP and power generation efficiency of the thermoelectric power generator. Both are in equilibrium. These curves depend on the performance of the devices and the converging temperature does not necessarily always satisfy $T_{j,\max}$. Hence, a careful optimization in the design is still required.

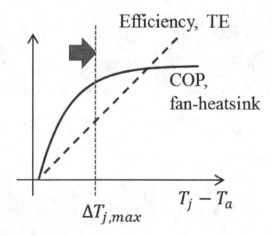

Fig. 2.9: General trend of cooling COP of fan−heat sink and the power generation efficiency of thermoelectric generator. This system successfully exists if the energy balance is satisfied at the temperature limit or lower: {Efficiency × COP} ≥ 1 at $T = (T_j - T_a) \leq \Delta T_{j,\max}$.

2.5   Thermoelectric Refrigeration

A wine cooler is a popular consumer product utilizing thermoelectric, while large-scale commercial wine coolers use a separated VC chiller. The advantage of thermoelectric is not only the low noise and low vibration, but also the COP for small temperature difference is good even when compared to the simplest VC with fixed-speed compressor. The detail is discussed in Chapter 6.

2.6   Thermal Control Equipment

The advantages of the thermoelectrics are not just in thermoenergy conversion, but this is also commonly recognized as the means for precise temperature control, reliable solid-state operations with no noise and no vibration, fully reversible in polarity, desirable flat and thin geometric shape for integration, and rapid response time for step change or high-frequency thermal cycling. Hence, the majority of the utilizations of thermoelectric modules is seen to be in electrical thermal control equipment. Scientific laboratory equipment with temperature control will likely have a thermoelectric temperature controller built-in, such as in a water chiller and incubator due to the requirement of low acoustic and mechanical noise and vibrations in contrast to VC cycles or Stirling refrigeration equipment.

2.7   Summary

Thermoelectric applications for photonics and electronics are reviewed and discussed. Thermoelectric solid-state cooling of the devices has advantages in vibration, acoustic noise, etc. compared to the vapor compression cycles or other mechanical thermodynamic cycles. In addition, the planner structure of thermoelectric module provides a capability of compact packaging for thermal management, refrigeration, or waste heat recovery. On the challenging side, the performance of thermoelectric heat energy conversion is generally moderate. Optimization is the key in order to maximize the cooling or power output. The optimization is especially important in waste heat recovery to minimize the initial cost. Although the thermoelectric energy conversion does not perform with as good efficiency as mechanical thermodynamic cycles in comparative scales, it shows better efficiency in smaller scale, e.g. less than 100 W. Mechanical thermodynamic cycles are based on volume change, whereas the heat transfer is a surface effect. Therefore, the thermoelectric thermal energy conversion technology is almost equally performing in a much smaller power range and is hence scalable. In another aspect, the precise temperature controllability and high-frequency response of the thermoelectric technology are suitable for the thermal management of electronic devices and therefore have been used. Small power thermal energy harvesting is a well-suited area for thermoelectrics. To conclude, it is critical to give a proper heat sink to match the best possible energy transport condition for thermal energy conversion with thermoelectric devices.

References

[1] A. Bar-Cohen (ed.), *Encyclopedia of Thermal Packaging: Thermal Packaging Technique*, World Scientific (2012).

[2] M. Iyengar, D. Milnes, P. Pritish, V. Kamath, B. Kochuparambil, D. Graybill, M. Schultz *et al.*, Server liquid cooling with chiller-less data center design to enable significant energy savings, *28th Annual IEEE Semiconductor Thermal Measurement and Management Symposium* (SEMI-THERM), 212–223 (2012).

[3] J. Koomey, Growth in data center electricity use 2005 to 2010, A report by Analytical Press, completed at the request of *The New York Times*, 9 (2011).

[4] M. Arlitt, C. Bash, S. Blagodurov, Y. Chen, T. Christian, D. Gmach, C. Hyser *et al.* Towards the design and operation of net-zero energy data centers, in 13th InterSociety Conference on Thermal and Thermomechanical Phenomena in Electronic Systems, *IEEE*, 552–561 (2012).

[5] C. B. Vining, An inconvenient truth about thermoelectrics, *Nat. Mater.*, **8**(2), 83 (2009).

[6] Y. Joshi, P. Kumar, eds. *Energy Efficient Thermal Management of Data Centers*, Springer Science & Business Media (2012).

[7] W. M. Kays, A. L. London, *Compact Heat Exchangers*, Third Edition, (1984).

[8] E. N. Sieder, G. E. Tate, Heat transfer and pressure drop of liquids in tubes, *Ind. Eng. Chem. Res.*, **28**(12), 1429–1435 (1936).

[9] K. Yazawa, A. Fedorov, Y. Joshi and A Shakouri, Energy efficient solid-state cooling for hot spot removal, in *Cooling of Microelectronic and Nanoelectronic Equipment: Advances and Emerging Research*, pp. 195–226 (2015).

[10] G. E. Moor, *Electron. Magaz.*, **38**(8), 4–6 (1965).

[11] P. Wang, Peng, On-chip thermoelectric cooling of semiconductor hot spot, PhD diss. (2007).

[12] X. Fan, G. Zeng, C. LaBounty, J. E. Bowers, E. Croke, C. C. Ahn, S. Huxtable, A. Majumdar and A. Shakouri, SiGeC/Si superlattice microcoolers, *Appl. Phys. Lett.*, **78**(11), 1580–1582 (2001).

[13] W. Liu and A. A. Balandin, Thermoelectric effects in wurtzite GaN and $Al_xGa_{1-x}N$ alloys, *J. Appl. Phys.*, **97**(12), 123705 (2005).

[14] J. Cho, Z. Li, E. Bozorg-Grayeli, T. Kodama, D. Francis, F. Ejeckam, F. Faili, M. Asheghi and K. E. Goodson, Improved thermal interfaces of GaN–diamond composite substrates for HEMT applications, *IEEE Trans. Compon. Packag. Manuf. Technol.*, **3**(1), 79–85 (2013).

[15] A. Bar-Cohen, J. J. Maurer and A. Sivananthan, Near-junction microfluidic cooling for wide bandgap devices, *MRS Adv.*, **1**(2), 181–195 (2016).

[16] Available at: https://www.photonics.com/Articles/Semiconductor_Lasers_An_Overview_of_Commercial/a25099.

[17] R. Beach, W. J. Benett, B. L. Freitas, D. Mundinger, B. J. Comaskey, R. W. Solarz and M. A. Emanuel, Modular microchannel cooled heatsinks for high average power laser diode arrays. *IEEE J. Quant. Electron.*, **28**(4), 966–976 (1992).

[18] D. R. Mundinger, R. Beach, W. Benett, R. Solarz, W. Krupke, R. Staver and D. Tuckerman, Demonstration of high-performance silicon microchannel heat exchangers for laser diode array cooling, *Appl. Phys. Lett.*, **53**(12), 1030–1032 (1988).

[19] J. D. Engerer and T. S. Fisher, Flash boiling from carbon foams for high-heat-flux transient cooling, *Appl. Phys. Lett.*, **109**, 024102 (2016).

[20] L. Shen, H. Chen, F. Xiao, Y. Yang and S. Wang, The step-change cooling performance of miniature thermoelectric module for pulse laser, *Energy Convers. Manag.*, **80**, 39–45 (2014).

[21] K. P. Pipe, R. J. Ram and A. Shakouri, Internal cooling in a semiconductor laser diode, *IEEE Photoni. Tech. Lett.*, **14**(4), 453–455 (2002).

[22] K. Yazawa, Thermal challenges in LED-driven display technologies: The early days, in *Thermal Management for LED Applications*, Eds. C. J. M. Lasance and A. Poppe, New York: Springer, pp. 477–498 (2014).

[23] G. J. Snyder, Application of the compatibility factor to the design of segmented and cascaded thermoelectric generators, *Appl. Phys. Lett.*, **84**(13), 2436–2438 (2004).

[24] Available: https://www.nasa.gov/sites/default/files/files/MMRTG.pdf.

[25] D. Bhatia, S. Bairagi, S. Goel and M. Jangra, Pacemakers charging using body energy, *J. Pharm. Bioallied Sci.*, **2**(1), 51 (2010).

[26] D. Bhatia, S. Bairagi, S. Goel, M. Jangra, Pacemakers charging using body energy, *J. Pharm. Bioallied Sci.*, **2**(1), 51–54 (2010).

[27] J. A. Paradiso, T. Starner, Energy scavenging for mobile and wireless electronics, *IEEE Pervasive Comput.*, **4**(1), 18–27 (2005).

[28] Beeby, S. P. and White, N. M., Low power systems (Section 5.4.1), in *Energy Harvesting for Autonomous Systems*, Artech House, 149–151 (2010).

[29] K. Ebrahimi, G. F. Jones and A. S. Fleischer, A review of data center cooling technology, operating conditions and the corresponding low-grade waste heat recovery opportunities, *Renew. Sust. Energ. Rev.*, **31**, 622–638 (2014).

[30] A. B. Little and S. Garimella, Comparative assessment of alternative cycles for waste heat recovery and upgrade, *Energy*, **36**(7), 4492–4504 (2011).

[31] K. Yazawa and A. Bar-Cohen, Method and apparatus for converting dissipated heat to work energy, U.S. Patent 6668550, issued December 30 (2003).

[32] X. Wu, F. Lai, L. Lin, Y. Li, L. Lin, Y. Qu and Z. Huang, Influence of thermal cycling on structural, optical and electrical properties of vanadium oxide thin films, *Appl. Surf. Sci.*, **255**(5), 2840–2844 (2008).

[33] K. Yazawa, G. L. Solbrekken and A. Bar-Cohen, Thermoelectric-powered convective cooling of microprocessors, *IEEE Trans. Adv. Packag.*, **28**(2), 231–239 (2005).

Chapter 3

Heat Transfer in Thermoelectric System

This chapter discusses important heat transfer mechanisms including major parasitic heat losses involved in thermoelectric energy conversion devices and systems. Due to the electrothermal coupling in the system, the analysis and modeling of fundamental heat transfer is indispensable for accurate performance optimization of a thermoelectric system assembled with heat sinks and thermal contacts. In addition, electrical parasitic losses are discussed as they cannot be separately determined from thermal contacts parasitic losses.

3.1 Heat Conduction in Thermoelectrics

Heat conduction follows the well-known Fourier law [1] and is a key principal mechanism determining the internal temperature gradient across the thermoelectric legs. Power generation depends on the temperature gradient applied. The heat conduction also contributes to the reverse heat flow in refrigeration applications. In the Fourier law, the amount of heat transported per unit cross-sectional area is linearly proportional to the temperature gradient. The heat current density, or heat flux, q_{cond} [W/m^2], is the key parameter of thermal energy harvesting. The general description of a one-dimensional (1D) heat flow is given by

$$q_{\mathrm{cond}} = -\beta \frac{dT}{dx} \tag{3.1}$$

where β is the thermal conductivity in $W/(m\,K)$. The negative sign on the right-hand side of the equation indicates that the heat flow is in the opposite direction to the temperature gradient. Temperature dependency of thermal conductivity takes effect when there is a large temperature gradient across the material of the thermoelectric element (leg). Also, β can be position-dependent in an inhomogeneous material. Thus, a more general expression becomes

$$q_{\mathrm{cond}} = -\beta(T, x)\frac{dT}{dx} \qquad (3.2)$$

3.1.1   *Heat spreading and contraction*

In a thermoelectric module, conductive heat transport not only happens as (1D) heat flow in the legs but also includes spreading and contraction through the substrates near the contacts with the legs. In most cases, the cross-section area a $[m^2]$ of a thermoelectric leg is smaller than the designated area A $[m^2]$ for each leg, defined as the total heat input area divided by the number of legs. This fractional area coverage is called the fill factor, $F = a/A$, which is a significant design parameter especially for cost-effective power generation. Details of the cost-effective design optimization is discussed in Chapter 5. Heat flow from the heat source comes into the hot side substrate and then to the leg through the thermal contact between the substrate and the leg. After going through the leg, the heat spreads again in the cold side substrate. Figure 3.1 illustrates the heat flow through the multi-layers and contacts around a TE leg. The substrates are typically made of ceramic due to its desirable electrical insulation, but there are some modules without substrates. In either case, thermoelectric legs must be mechanically held strong by the structure. Electrodes connect the p-type and n-type semiconductor TE legs in series and make the thermal contact to the hot or cold side as well.

In a thermoelectric module, a pair of p-type and n-type legs are typically taken as the unit of the model. Then, the performance is simply scaled with the number of legs, N_{pair}. For simplicity, the cross-sectional area of p-type and n-type legs are assumed to be the same,

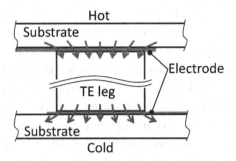

Fig. 3.1: Heat flow contraction at the hot side and spreading at the cold side of a TE leg. Thermal resistance due to the contraction/spreading depends on the fractional area coverage F of the leg, the thickness of the substrate, and the thermal conductivity.

being denoted as A_{leg}, and then fill factor F is given by

$$F = 2N_{\text{pair}}A_{\text{leg}}/A_{\text{module}} \qquad (3.3)$$

where A_{module} is the total footprint of the module, i.e. total heat input area.

An off-the-shelf product typically has a relatively large fill factor ranging 0.4–0.6, where the design is likely intended to collect the heat flow into the legs as much as possible. This way, the heat input is close to the maximum amount. However, due to the small aspect ratio of legs, i.e. length-to-area ratio, the temperature difference applied across the leg becomes relatively small compared to those across the thermal contacts. In fact, the optimal design for maximizing power output is that the temperature difference across the TE leg is approximately a half of the total temperature difference applied end to end (see Section 5.2 in Chapter 5). Thus, a large fill factor may require longer TE legs to keep the temperature gradient near optimal, which in turn requires more material usage than the cases of a lower fill factor.

A smaller fill factor requires a smaller mass of material and thus saves the use of expensive thermoelectric materials [2]. As the fill factor gets further smaller by reducing the leg area, the interfacial thermal resistances including the contraction and spreading effects may rapidly increase, and thus, the temperature gradient across the leg is adversely reduced. Hence, there is an optimum for the fill factor, which is depending on the heat flux conditions. The minimum cost per power is typically in the range between 0.1 and 0.2 \$/W [3]. Some recent thermoelectric modules use relatively smaller fill factors, e.g. 25–30%. For such smaller fill factors, contraction/spreading thermal resistances need to be carefully considered in the module design.

Spreading and contraction thermal resistances are essentially symmetric in the hot and cold sides of module and just temperature gradients are in the opposite direction. Hence, a heat transfer model with a single formula can be utilized for both spreading and contraction.

The spreading thermal resistance in a rectangular system with compound substrates is well summarized in the work by Muzychka *et al.* [4]. The thermal resistance of a square substrate that includes the spreading effect in contact with a smaller TE leg at the center of the substrate is given by

$$\psi_s = \frac{1}{2a^2b^2\beta} \sum_{m=1}^{\infty} \frac{\sin^2(a\delta_m)}{\delta_m^3} \cdot \varphi(\delta_m) + \frac{1}{2a^2b^2\beta} \sum_{n=1}^{\infty} \frac{\sin^2(a\lambda_n)}{\lambda_n^3} \cdot \varphi(\lambda_n)$$

$$+ \frac{1}{2a^4b^2\beta} \sum_{m=1}^{\infty} \sum_{n=1}^{\infty} \frac{\sin^2(a\delta_m)\sin^2(a\lambda_n)}{\delta_m^2\lambda_n^2\gamma_{m,n}} \cdot \varphi(\gamma_{m,n}) \qquad (3.4)$$

where b[m] and d_{sub}[m] are, respectively, the side length and the thickness of the square substrate, β_{sub} [W/(m K)] is the thermal conductivity of the substrate, h[W/(m^2 K)] is the heat transfer coefficient at the back side of the substrate, and a [m] is the side length of the contact ($a < b$). Also,

$$\varphi(\xi) = \frac{\left(e^{2\xi d_{\text{sub}}} + 1\right)\xi d_{\text{sub}} - \left(1 - e^{2\xi d_{\text{sub}}}\right)h d_{\text{sub}}/\beta}{\left(e^{2\xi d_{\text{sub}}} - 1\right)\xi d_{\text{sub}} + \left(1 - e^{2\xi d_{\text{sub}}}\right)h d_{\text{sub}}/\beta} \qquad (3.5)$$

where ξ is a variable representing the respective eigenvalue. The eigenvalues for these solutions are $\delta_m = m\pi/b$, $\lambda_n = n\pi/b$, and $\gamma_{m,n} = \sqrt{\delta_m^2 + \lambda_n^2}$. Note that the above model is based on the isoflux condition in a rigorous discussion. The isoflux assumption is reasonable for passive heat conduction, e.g. in electronics cooling applications, but not always valid for the high heat transfer rate convection heat sinks with liquid flow, e.g. microchannels. For such cases, a full 3D numerical simulation is recommended.

3.1.2 Closed formula

As long as the fill factor is not smaller than 0.05, the closed formula developed by Song *et al.* [5] is useful with an engineering accuracy. The model is applicable to square plates, while it was developed based on their earlier investigation for the variation of the contacting geometries [6]. The closed-form solution of the spreading resistance is derived in two separate forms, which are, respectively, the average and the maximum of the thermal resistance given by

$$\psi_{\text{ave}} = \frac{1}{2} (1 - \varepsilon)^{3/2} \Phi_c \qquad (3.6)$$

$$\psi_{\text{max}} = \frac{1}{\sqrt{\pi}} (1 - \varepsilon) \Phi_c \qquad (3.7)$$

where

$$\Phi_c = \frac{tanh\,(\lambda_c \tau) + \frac{\lambda_c}{\text{Bi}}}{1 + \frac{\lambda_c}{\text{Bi}} tanh\,(\lambda_c \tau)} \qquad (3.8)$$

with

$$\lambda_c = \pi + \frac{1}{\varepsilon \sqrt{\pi}} \qquad (3.9)$$

and $\varepsilon = a/b = \sqrt{A_{\text{leg}}/\pi}/\sqrt{A_{\text{sub}}/\pi}$, $(a < b)$ is the ratio of the effective diameters of the areas. Subscripts *leg* and *sub* denote the quantities of the leg and the substrate, respectively. The dimensionless thickness of the substrate is defined as

$$\tau = d_{\text{sub}}/b = d_{\text{sub}}/\sqrt{A_{\text{sub}}/\pi} \qquad (3.10)$$

The Biot number is defined as $\text{Bi} = hb/\beta_{\text{sub}}$. In thermoelectric applications, this approach of averaging thermal resistance works well. The literature claimed the errors are within 10% over the range compared to the analytical solution. For a thermoelectric power generation case, this accuracy applies for fill factors in the range of 0.05–0.833.

3.2   Forced Convection Heat Exchange via Fluid

This section deals with convective heat transfer from/to the thermo-electric module via fluid flow through a heat sink at least on one side or both sides. From the view of thermoelectric modules, an external heat reservoir is connected to the legs through a fluid—solid convection heat transfer. This happens in a typical condition of power generation, heat recovery, or heat pumping. Without refreshing the heat-carrying fluid on either hot or cold side, the temperature at the terminal could not be maintained constant in real systems. Without continuous mass flow, the concept of temperature reservoir may not sustain over time. Therefore, the consideration of convective heat transport in thermoelectric refrigeration and power generation systems may be critical.

3.2.1   *Case 1: Solid heat source and cooling by fluid*

Here, we first discuss the case of a solid heat source that transfers heat to the TE module by conduction and a convective heat sink at the cold side. This case is similar to an electronics or photonics cooling with a cold plate or a combined heatsink. The only difference is the thermoelectric placed in between the two. A typical example is shown in Figs. 3.2 and 3.3. Cold plate allows water or other working fluid to go through and remove heat from the TE module using sensible heat capacity in a single phase. When the channel diameter gets smaller, the removed heat flux per given flowrate of working fluid increases, so that mini- or microchannels [7] show better overall heat transfer as long as a sufficient pump work is available. A two-phase microchannel heat sink can be utilized as well to further enhance the heat transfer with latent heat [8].

The heat is supplied from a solid heat source. This configuration is possible either for thermoelectric power generation or thermoelectric active cooling. Depending on power generation or cooling, however, the amount of heat required to be rejected through the cold plate varies. Typically, a cooling system dissipates a larger amount of heat than a power generation since the Joule heat that occurs in the cooling process is added to the heat from the source.

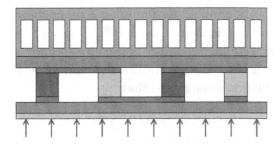

Fig. 3.2: Cold plate configuration. The bottom plate is in contact with a heat source. At the top side is a microchannel heat sink or a cold plate that removes heat from the multi-leg TE module below. In this case, the heat source and the inlet fluid are considered as the temperature reservoirs.

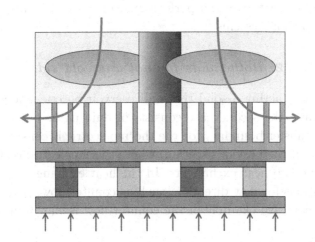

Fig. 3.3: Fan-based air heatsink configuration. The bottom plate is a heat source in contact with the thermoelectric module. At the top side is a cooling fan. The heat source and the inlet air are considered as the temperature reservoirs.

Air convection heat removal is more popular, as shown in Fig. 3.3. Typically, air at room temperature is used as the coolant. This configuration also fits both power generation and thermoelectric active cooling. Heat amount rejected through the heat sink is different for this case as well.

The most notable difference between the two configurations shown in Figs. 3.2 and 3.3 is found in the energy efficiency of pumping the refrigerant flow as a function of heat flux. In a laminar flow, the amount of work required for pumping is proportional to the third power of the volumetric mean flow velocity $\bar{u} = \dot{V}/S$, where $\dot{V}$ is the volumetric flow rate [m³/s] and S is the cross-section area [m²] perpendicular to the flow direction. The dynamic pressure loss ΔP is quadratic to the flow velocity, so that

$$W_{pp} = \Delta P \dot{V} \propto \bar{u}^3 \tag{3.11}$$

where ΔP is the pressure drop, which is a function of viscosity. Considering density, viscosity, specific heat, and thermal conductivity, water has a greater capacity than air in removing heat per unit input work. This difference between the heat transfer solutions is discussed more in detail for the overall performance of trade-off analysis in TE power generation systems [9].

3.2.2 *Case 2: Both fluids for hot and cold sides*

Instead of contacting a solid thermal reservoir on one side as in the previous case, the second case consists of fluid paths on both sides. The heat transfer boundary for the thermoelectric is considered similar to that in the previous section, where thermoelectric is placed in between the two heat exchangers. In the heat exchangers, there are several options for flow direction, such as counter flow, parallel flow, and cross flow. An example of the counter-flow is shown in Fig. 3.4.

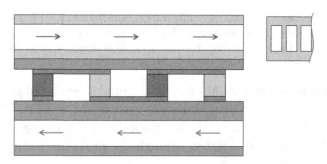

Fig. 3.4: An example of the counter-flow configuration for two-fluid flow. Both hot and cold side channels can be microchannel or just open channels. A parallel flow configuration utilizes two flows in the opposite directions. A cross-flow system consists of two flows in perpendicular directions.

In a power generator system, the temperature of a hot fluid is the source of energy, and that of a colder fluid is used to remove heat from the other side of the TE module. A temperature difference across the thermoelectric legs is thus generated by the temperature difference between the two fluids.

In a room-temperature cooling system, the temperature reduction of cooling fluid flow is an important indication of performance. In this case, thermoelectric works as heat pump — heat absorbed from the colder flow is pumped to the warmer flow by applying the electrical power into the module. In another usage of thermoelectric heat pumping, a heating mode rather than a cooling mode is also available. By pumping heat from the cold fluid to the hot fluid, the thermoelectric heating is more efficient than an electrical resistive heater since the internal heat generation is added on the pumping heat.

3.2.3 *Effective heat transfer coefficient*

Convective heat transport from the channel wall to the working fluid is only a part of the heat removal process. The sensible heat removal by mass transfer of the fluid also has an impact on the heat removal performance. A higher thermal resistance from either one of them would dominate the thermal performance.

Here, the effective heat transfer coefficient at the footprint area is not necessarily equal to that of the channel wall due to the enlarged surface area provided by the channels. Figure 3.5 shows an example of rectangular duct channels attached to TE modules at both sides.

Given the condition that the material thermal conductivity of channels is sufficiently large, the heat transfer is boosted by the factor of the area ratio of the channel surface area to the module area under the assumption of uniform convection heat transfer over the channel walls. In the symmetric module stack shown in Fig. 3.5, the heat transfer coefficient at the interface between the channels and the thermoelectric module is then given by

$$h_{TE} = \frac{(a+b)}{c} h_{ch} \qquad (3.12)$$

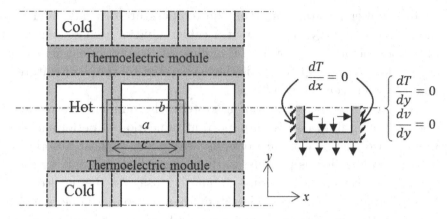

Fig. 3.5:   Cross-section of the channel/TE stack (left) and the details of half-channel (right). Around the center line, the velocity and temperature profiles are symmetric.

The channel heat transfer coefficient h_{ch} can be found from the Nusselt number depending on the fluid and channel conditions, as explained in the following sections.

3.2.3.1   *Fully developed laminar flow*

In thermoelectric applications, the heat amount brought in by the fluid is a significant factor for cost effectiveness. In the case of liquid as a working fluid, the flow regime is likely to be laminar, whereas it is more likely to be turbulent for gases and air.

At the fully developed laminar flow, the Nusselt number is constant independent of the Reynolds number (based on the mean fluid velocity). The constant Nusselt number depends on the geometrical factor (see Kays and London [10]). As an example, the Nusselt number is 3.61 for a square channel with circumferentially uniform heat flux (isoflux) condition. Shah and London investigated some variation of the cross-section geometries and thermal boundary conditions [11]. The heat transfer coefficient at the effective wall is then given by

$$h_{ch} = \frac{\beta_f}{D_h} \mathrm{Nu} \qquad (3.13)$$

where β_f is the thermal conductivity of the fluid and D_h is the hydraulic diameter, e.g. $D_h = w$, the side length of a square duct. In practice, the thermal boundary entrance length (developing region) depends on the channel design and inlet conditions. Hence, the developing region can affect the average heat transfer coefficient for a short channel, as discussed in the following section.

3.2.3.2 *Developing laminar, transitional, and turbulent flows*

For a laminar flow, the average Nusselt number must include those of the entry region and the fully developed region and is given by Sieder and Tate [12] as

$$\mathrm{Nu} = 1.86 \left(\frac{\mathrm{RePr}}{L/D_h} \right)^{1/3} \left(\frac{\mu}{\mu_s} \right)^{0.14} \tag{3.14}$$

where L is the channel length. This correlation takes into account the change in the dynamic viscosity from μ at the bulk medium temperature to that of the wall surface μ_s due to the temperature change from the mean fluid to the wall, respectively. This correlation is valid for the fluid with $0.6 \leq \mathrm{Pr} \leq 5$ and when $0.0044 \leq (\mu/\mu_s) \leq 9.75$. Note that deionized water under atmospheric pressure or typical refrigerant such as R-22 under compression likely exceeds the valid range of Prandtl number for the above correlation as $\mathrm{Pr} > 5$. For such cases, the hydraulic boundary develops quickly where the entrance effect is rather small, hence the following correlation for fully developed flow by Kays and London [9] is recommended:

$$\mathrm{Nu} = 3.66 + \frac{0.0668 \, (D_h/L) \, \mathrm{RePr}}{1 + 0.04 \left[(D_h/L) \, \mathrm{RePr} \right]^{2/3}} \quad \text{for } \mathrm{Pr} > 5 \tag{3.15}$$

In typical microchannels, the length of the channel is much larger than the hydraulic diameter. Hence, it is reasonable to estimate the performance based on the 'fully developed flow'. More details can be found in Ref. [13].

Above correlations are based on circular tubes and isothermal (uniform temperature) condition and are different for different channel geometry thermal boundary conditions (see Ref. [14] for details).

For a turbulent flow, Sieder and Tate [12] also gave the following correlation, which includes the temperature-dependent viscosity change between the mean fluid and the nearby wall:

$$\mathrm{Nu} = 0.027\mathrm{Re}^{4/5}\mathrm{Pr}^{1/3} \left(\frac{\mu}{\mu_s}\right)^{0.14} \tag{3.16}$$

This correlation is valid for Re $\geq$ 10,000 the fluid 0.7 $\leq$ Pr and the channel design with $L/D_h \geq 10$. Although the transitional point between laminar and turbulent regimes is not clearly determined, the point of drastic change in the friction plot suggests a reasonable range Re $\leq$ 2,100 for laminar flows (Eq. (3.14)) and Re $\geq$ 2,100 for turbulent flows (Eq. (3.16)).

3.2.4   *Effectiveness of thermoelectric integrated heat exchanger*

Due to the flat interfaces, a typical TE module has the potential to be integrated with heat exchangers and developed into a thermoelectric-embedded plate heat exchanger system. This unique configuration, which has not yet been commercialized on the market, would have the advantage of switching operations from power generation to active heat pumping in either direction in addition to the passive heat diffusion.

A variety of flow configurations in the heat exchangers are possible, including parallel flows, counter flows, and cross flows, as discussed above. The parallel- and counter-flow configurations have been compared for thermoelectric power generation in Ref. [14], but other than that, there are not many published results so far. Depending on the flow configuration, the temperature profile across the lateral direction of the thermoelectric module can be non-uniform. Such a 2D temperature profile can potentially impact the overall performance, but it has been considered only as the secondary effect or omitted due to the complexity. Since it is too complex to bring to the analytical model, numerical calculations or hybrid with an analytical model for a single leg will be a practical method to determine the local temperature gradients and total performance with the double integral along the two axes of the coordinate.

In a typical counter-flow configuration, the performance of heat transfer is represented by the effectiveness ε in a two-fluid system, which measures the capability of heat transfer across the channels from one fluid flow to another in comparison to the maximum heat amount that can be transferred, such that

$$\varepsilon = \frac{Q_{\text{transfer}}}{Q_{\text{max}}}, \quad 0 \leq \varepsilon \leq 1 \tag{3.17}$$

If the fluid temperatures are given, the effectiveness is found as

$$\varepsilon = \frac{T_{h,\text{in}} - T_{h,\text{out}}}{T_{h,\text{in}} - T_{c,\text{in}}} \tag{3.18}$$

$$= \frac{T_{c,\text{out}} - T_{c,\text{in}}}{T_{h,\text{in}} - T_{c,\text{in}}} \cdot \frac{C_c}{C_h} \tag{3.19}$$

where C is the heat capacitance given by $C_h = \dot{m}_h c_{p,h}$, $C_c = \dot{m}_c c_{p,c}$, respectively, in the unit of [W/K]. The subscript 'h' indicates the hot side, and the subscript 'c' indicates the cold side. $\dot{m}$ is the mass flow rate in [kg/s] and c_p is the specific heat in [J/(kg K)]. If the both fluid flows are in the same closed system, C_h and C_c are equal to each other due to the mass conservation.

When a thermoelectric power generator is integrated with a heat exchanger, the first-order performance can be quickly calculated by using the mean temperature difference ΔT_m across the flow separation plate (thermoelectric and the hot and cold side plates in the heat exchanger. This temperature difference is used as that between the source and thermal ground. This temperature is known as the log mean temperature difference (LMTD) and is found as

$$\Delta T_m = \frac{(T_{h,\text{in}} - T_{c,\text{out}}) - (T_{h,\text{out}} - T_{c,\text{in}})}{\ln\left((T_{h,\text{in}} - T_{c,\text{out}}) / (T_{h,\text{out}} - T_{c,\text{in}})\right)} \tag{3.20}$$

This is based on the exponential nature of the temperature profile along the flow direction as the heat exchange happens perpendicularly across the plate.

In the design phase of the heat exchanger, the outlet temperatures of both the hot and cold flow channels may not be known. In a counter-flow configuration, the effectiveness can also be obtained using the number of transfer units (NTU) method as

$$\varepsilon_{counter\text{-}flow} = \frac{1 - e^{-\text{NTU}(1-\alpha)}}{1 - \alpha e^{-\text{NTU}(1-\alpha)}} \tag{3.21}$$

where $\alpha = \frac{C_{\min}}{C_{\max}}$, $C_{\min} = |C_h, C_c|_{\min}$, $C_{\max} = |C_h, C_c|_{\max}$, and C_h and C_c are the heat capacity rates at the hot and cold sides, respectively. $\text{NTU} = UA/C_{\min}$, with U being the overall heat transfer coefficient, and A is the heat transfer area. Thus, the product UA is the thermal conductance [W/K] across the plate between the channels, typically described as

$$UA = A / \{1/h_h + d/\beta_{\text{plate}} + 1/h_c\} \tag{3.22}$$

This formula is based on the approximation that the lateral thermal conduction is sufficiently small at the area of contact between the flow and the flow separation plate. Practically, it expects low thermal conductivity (β_{plate}) in a lateral direction of a very thin (thickness d) intermediate plate.

For a parallel flow, the effectiveness is determined as

$$\varepsilon_{parallel\_flow} = \frac{1 - e^{-\text{NTU}(1+\alpha)}}{1 + \alpha} \tag{3.23}$$

For further variation, Kays and London [9] delivered more details.

3.2.5   *Pump power and cooling efficiency*

In both cases of power generation and heat pump, the pump power W_{pp} for a fluid through heat exchanger is a factor that must be taken into account to determine the net efficiency or net COP of energy conversion. The pump power as a fluid dynamic work shall match or be larger than the value of {pressure head} × {volumetric flow rate}. The efficiency of a fluid power driven by an electrical power needs to be considered. The efficiency of a water pump is usually around 90% and that of a fan motor is typically ranging 25–60% depending on the size, power scale, and the casing of fluid dynamic design. Then the net power output and the net coefficient of performance of a

TE system are given, respectively, by

$$W_{net.generation} = W_{\text{TEG}} - W_{\text{pp}}/\eta_{\text{pp}} \qquad (3.24)$$

$$\text{COP}_{net.cooling} = \frac{Q}{W_{\text{TEG}} + W_{\text{pp}}/\eta_{\text{pp}}} \qquad (3.25)$$

Pressure head ΔP is found by using the Darcy–Weisbach equation [15] through the knowledge of the friction factor f for the channel(s) from the geometry and dimensions. The mean flow velocity v is found from the volumetric flow rate $\dot{V} = vnA$, where A is the cross-section area and n is the number of channels in parallel. For a serpentine channel, n is usually unity and the channel length is much larger than the size of plate:

$$W_{\text{pp}} = \Delta P \dot{V} \qquad (3.26)$$

$$\Delta P = f \frac{\rho v^2}{2} \frac{L}{D_h} \qquad (3.27)$$

where L is the channel length, ρ is the density of the fluid, and D_h is the hydraulic diameter defined as

$$D_h = 4 \frac{(\text{Cross-sectional area})}{(\text{Wetted perimeter})} \qquad (3.28)$$

3.3 Radiation Heat Energy Exchange

Unlike other heat engines, a thermoelectric system is well fit to use radiation heat transfer as the heat transfer mechanism either for hot (heat source) or cold (dissipation) side for power generation applications. This heat transfer is independent of the convective heat transfer as long as the system is in equilibrium. Radiation heat transfer follows the gray body Stefan–Boltzmann law given by

$$E = \varepsilon \sigma_{SB} T^4 \qquad (3.29)$$

where E is the emissive power — energy emitted per unit area per unit time in [W/m^2], T is the absolute surface temperature, and

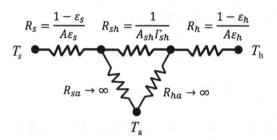

Fig. 3.6: Radiation heat transfer network between two parallel flat plate and ambient air. With good insulation, R_{sa} and R_{ha} are large, so that heat transfer through these resistances can be neglected.

σ_{SB} [W/(m^2 K^4)] is the Stefan−Boltzmann constant given by

$$\sigma_{SB} = \frac{2\pi^5 k^4}{15 c^2 h^3} = 5.67 \times 10^{-8} \, [\text{W}/\text{m}^2\text{K}^4] \tag{3.30}$$

where k is the Boltzmann constant, c is the speed of light, and h is the Prank constant. ε is the average emissivity of the surface, which is the optical property of the surface ($\varepsilon < 1$ for a gray surface).

Between two temperatures, e.g. source temperature T_s and the hot side temperature of thermoelectric T_h, the total radiative heat flux q [W/m^2] is found as

$$q = \sigma_{SB} \left(T_s^4 - T_h^4 \right) / \left\{ \frac{1 - \varepsilon_s}{\varepsilon_s} + \frac{1}{\Gamma_{sh}} + \frac{1 - \varepsilon_h}{\varepsilon_h} \right\} \tag{3.31}$$

where Γ is the view factor between the two surfaces and the denominator of the above equation is found from the radiative heat transfer network analysis between two flat surfaces, as shown in Fig. 3.6. In this equation, the cross-talk of the ambient temperature is neglected by assuming the insulation around the edges. The view factor is determined by the geometries of the two surfaces.

3.4   Electrothermal Contacts and Parasitic Losses

3.4.1   *Thermal bypasses*

As the area covered by thermoelectric legs is smaller than the substrate area or the overall footprint, the remaining gap between legs must be filled with air, vacuum space, or another material. Bypass heat flow through the gap space must be taken into account since this parasitic heat flow (cross-talk heat flow between the hot and cold sides) only reduces the performance. Due to its narrow space, regardless of the material used, whether gas, liquid, gel, or solid, the internal mass transfer is expected to be negligibly small. This means there is no convective heat transfer through the filler.

Since the gas that fills the gap is assumed to be stationary, the heat transfer by this gaseous filler is only through thermal conduction. However, the radiation heat exchange through the gas can still happen as long as the gas is transparent enough for the wavelengths of the radiation heat exchange. Yet, typically a large temperature difference more than several hundreds of Kelvin is required for the radiation heat transfer to be significant. The emissivity of the inner surface at infrared wavelengths is then a design parameter to minimize the bypass heat flow through the filler (Fig. 3.7).

The heat conduction and radiation bypass heat leaks through the filler are formulated in the following equations, respectively:

$$\psi_{TE\text{gap,cond}} = \frac{d_{TE}}{\beta_{\text{gap}}(1 - F)} \tag{3.32}$$

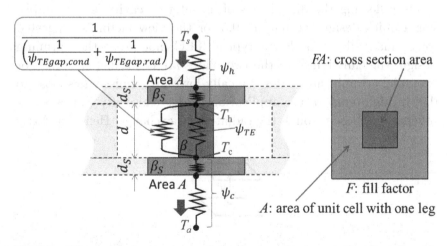

Fig. 3.7: Thermal resistance network for the parasitic bypass heat flow through fillers between TE legs.

$$\psi_{TE\text{gap,rad}} = \left\{ \frac{1 - \varepsilon_h}{\varepsilon_h} + \frac{1}{\Gamma_{hc}} + \frac{1 - \varepsilon_c}{\varepsilon_c} \right\} \frac{(T_h - T_c)}{\sigma_{SB} \left(1 - F \right) \left(T_h^4 - T_c^4 \right)}$$

$$(3.33)$$

where d [m] is the thickness of the thermoelectric element, β_{gap} [W/(m K)] is the thermal conductivity of the gap filler material, σ is the Stefan−Boltzmann constant, Γ_{h-c} is the radiative view factor between the hot and cold surfaces, which is in range of 0.2−0.5. As an example, the hot and cold surfaces between TE legs can be approximately described as equi-area rectangular parallel plates, as shown in Fig. 3.8.

The view factor is then found as

$$\Gamma = \frac{1}{\pi xy} \left[ln \frac{X^2 Y^2}{X^2 + Y^2 - 1} + 2x \left(Y \arctan \frac{x}{Y} - \arctan x \right) \right.$$

$$\left. + 2y \left(X \arctan \frac{y}{X} - \arctan y \right) \right]$$

$$(3.34)$$

where $X = \sqrt{1 + x^2}$ and $Y = \sqrt{1 + y^2}$, with $x = W_1/H$ and $y = W_2/H$. If $H = W_1 = W_2$, $\Gamma = 0.1998$. In the thermoelectric module, $H = d$, and $W_1 = W_2 = (1 - F) A$.

There are several studies in the literature, e.g. [16, 17], but accurately reflecting the dimensions of an internal cavity is not simple. For a quick design estimation, 0.5 for the view factor is suggested for a small fill factor. For a typical TEG module at the optimum design, the emissivities of the two plates are equal, i.e. $\varepsilon_h = \varepsilon_c$. For example, ceramic substrates typically have high emissivities close to 0.9. In high-temperature applications, large emissivity causes a significant heat loss through the radiate heat exchange. Hence, a glossy

Fig. 3.8:   Radiation heat transfer between equal rectangular parallel plates.

metal surface is highly desired to reduce the radiation. An extension of gold electrodes may give the emissivity of ~0.1 or even, which can lower the heat leak an order of magnitude smaller compared to bare ceramic surfaces.

The pressure dependency of the effective thermal conductivity of dry air obeys the following relationship according to [18]:

$$\beta_{air} = \beta_{air,atm} \left(\frac{1}{1 + \frac{7.6 \times 10^{-5}}{Pd/T}} \right) \tag{3.35}$$

where $\beta_{air,atm}$ is the thermal conductivity of dry air at the atmosphere pressure, P is the pressure [Torr], and d is the distance between the gap (leg length in our case), and T is the mean temperature of air. Assuming that the thermal conductivity of the TE material is 1.5 W/(m K) at the mean temperature of the application, with fill factor $F = 0.05$, and gold surface coating with emissivity of 0.02, e.g. an air pressure lower than 10 Torr is required to achieve the effective thermal conductivity as low as 0.004 W/(m K). Hence, vacuum packaging is highly desired [19] to suppress the heat leakage through the gap. Figure 3.9 shows the effective thermal conductivity of dry air as a function of pressure.

Another option for the gap filler is aerogel [20], in particular for high-temperature power generation applications, which could lower the conduction heat loss with a thermal conductivity even lower than that of air gaps and eliminate the thermal radiation cross-talk simultaneously.

The three components of heat transfer involved in the thermoelectric module, which are conduction through legs, conduction and radiation through gap fillers, are all a function of geometry. Figure 3.10 shows the relation as a function of fill factor as an example case of power generation with a material of $ZT = 1$. Temperatures of the hot and cold sides are assumed to be 600 K and 300 K, respectively. As shown in the figure, modules with low fill factors below 0.1 can have a significant impact due to the parasitic conduction through air gaps. Radiation loss can also amount to about 5% of the heat conduction through legs in this temperature range.

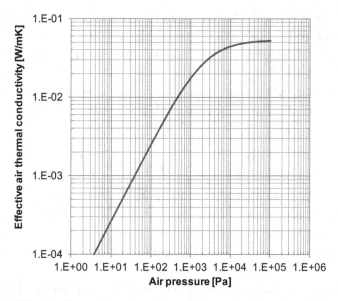

Fig. 3.9: Effective thermal conductivity of air gap as function of air pressure. Moderate vacuum allows several orders of magnitude lower heat conduction, which is desirable for high-temperature applications.

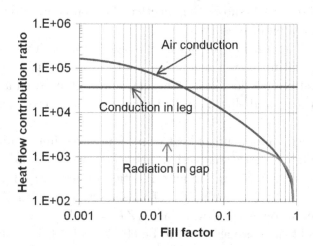

Fig. 3.10: Normalized heat flow contribution of three heat transfer components in power generation mode as functions of fill factor (design parameter). Temperatures of the hot and cold sides are 600 K and 300 K, respectively.

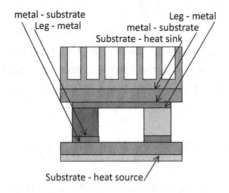

Fig. 3.11:   Contact interfaces in thermoelectric modules.

3.4.2   *Thermal interfaces*

There are several thermal interfaces along the heat flow in thermoelectric modules that can potentially impose significant thermal resistances. See Fig. 3.11 for the various interfaces existing in a TE system.

Typically, these interfaces are made through soldering or bonding processes, so that the contact resistances are relatively small compared to the 1D conduction through the leg and the spreading/contraction thermal resistances.

Another non-negligible interface of thermal resistance exists at the contact between the thermoelectric module and the heat sink on either side. The contact is typically made by applying pressure with a thermal paste. The method is similar to the thermal interface made for an electronic device cooling with a heat sink. For either direct refrigeration or power generation, both sides of the thermoelectric module have to have thermal interface resistances. Also, the temperature range is usually different from that of electronic device cooling. Selection of proper thermal interface material is important with an attention to the temperature range of the interface.

3.4.3   *Electrical parasitic losses*

Electrical parasitic losses are observed at any interface that current flows through. Although these are not the phenomena of heat transfer, electrical losses eventually create Joule heat when the system is in operation with electric current flow, which can reduce the power output in generation mode or the COP or cooling capacity in refrigeration. By taking into account the interfacial electrical resistance per contact R_{IF}, the total internal module resistance R_m becomes

$$R_m = N \left(2R_{IF} + \frac{d}{\sigma F A} \right) \qquad (3.36)$$

where N is the total number of legs and A is the total area of the module in [m^2]. The interfacial resistance R_{IF} consists of the contact resistance R_k that includes the current crowding and the additional series resistance R_s along the length of the metal electrodes between two adjacent legs:

$$R_{IF} = R_k + R_s \qquad (3.37)$$

In order to determine the contact resistance R_k, the transfer length L_T is introduced, which is the lateral distance that most of the current ($1/e$ decay) travels to reach the semiconductor solid underneath the metal contact, considering the exponentially decaying current distribution with distance based on Berger [21]:

$$L_T = \sqrt{\rho_c / R_{sh}} \qquad (3.38)$$

where ρ_c is the specific contact resistivity [Ω cm^2] of the contact interface between the leg and the metallization, which depends on both materials and the process of contact. Typically, a reasonably good contact has a contact resistance of 10^{-6}–10^{-10} [Ω cm^2] and R_{sh} is the sheet resistance [Ω] given by

$$R_{sh} = \rho_m / d_m \qquad (3.39)$$

where ρ_m [Ω m] is the resistivity of the metal and d_m [m] is the thickness of the metal electrode. This transfer length should cover the entire contact area of the leg, so that the whole area allows current to go through. Thus,

$$L_T \geq \sqrt{F/N} \tag{3.40}$$

The minimum metallization thickness is found as

$$d_{m\_min} = \frac{\rho_m}{\rho_c}\frac{F}{N} \tag{3.41}$$

According to a simplified two-dimensional model by Ono *et al.* [22], the contact electrical resistance R_k of the leg including the impact of metallization thickness is found as

$$R_k = \frac{\rho_c + \sqrt{\rho_c R_{sh}}W_1\coth\left(W/L_T\right) + 0.5R_{sh}W_1^2 \\ + \sqrt{\rho_c R_{sh}}W_2/\sinh\left(W/L_T\right)}{(W+W_1+W_2)\,L} \tag{3.42}$$

with $L = W = \sqrt{F/N}$ for a single leg (Fig. 3.12) and

$$L_1 = L_2 = W_1 = W_2 = \left(1 - \sqrt{F}\right)/2\sqrt{N} \tag{3.43}$$

Another parasitic resistance is the interconnect series resistance R_s due to the longer distance between legs, especially for the case of smaller fill factor F. The series resistance is defined along the distance between two adjacent legs. The series resistance per leg is

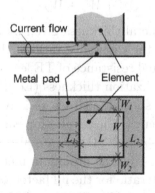

Fig. 3.12: Electric interfacial parasitic resistances. The figure shows only one contact and there are two contacts per leg in a module.

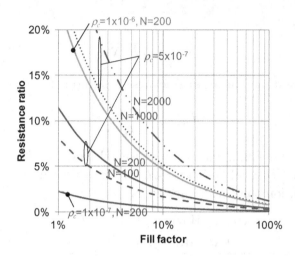

Fig. 3.13: Relative parasitic interfacial resistance as a function of fill factor. The resistance is normalized by that of the thermoelectric legs. Around 5% of relative parasitic resistance may be acceptable. The solid curves show variations of specific contact resistivity $1 \times 10^{-7}, 5 \times 10^{-7}$, and $1 \times 10^{-6} \, \Omega \, cm^2$ for the element area density $N = 200 \, (1/cm^2)$. The electrical conductivity of the metallization layer is $1.68 \times 10^{-8} \, \Omega \, m$ and the electrical conductivity of leg is assumed to be $\sigma_{TE} = 3.72 \times 10^4 \, (1/(\Omega \, m))$. Dashed curves show the variation of element density $N = 100, 200, 1000$, and 2000 for the contact resistivity of $5 \times 10^{-7} \Omega \, cm^2$.

found as

$$R_s = \rho_m \frac{L_1}{d_m \, (W + W_1 + W_2)} = \rho_c \frac{1 - \sqrt{F}}{F} N \qquad (3.44)$$

where the minimum metal thickness given by Eq. (3.41) was used.

Figure 3.13 shows the resistance ratio of the parasitic interfacial resistance to the internal resistance of TE legs with copper electrodes at the minimum metallization thickness (Eq. (3.41)) as a function of fill factor for several specific contact resistivities of $1 \times 10^{-7}, 5 \times 10^{-7}$, and $1 \times 10^{-6} \Omega \, cm^2$. The results for various element densities for specific resistivity $5 \times 10^{-7} \Omega \, cm^2$ are also displayed to show the impact of element density. The minimum metallization thickness to match to 5% resistance ratio for the fill factor varies depending on the specific contact resistivity. Metallization thickness linearly decreases as fill factor decreases, as shown in Eq. (3.41). For example, a 7.6 μm thick copper is enough to meet the 5% parasitic resistance ratio for the specific contact resistivity of $1 \times 10^{-6} \Omega \, cm^2$. Similarly, 5.4 μm for $5 \times 10^{-7}\,\Omega\,\mathrm{cm}^2$ and 3.0 μm for $1 \times 10^{-6}\,\Omega\,\mathrm{cm}^2$ are the minimum thicknesses for the metallization layers while the metal is copper. Thicker metallization increases the material use and its cost in the module. In the above case, the metallization cost is less than 34% of the leg cost for more than 1% of fill factor.

3.5 Passive Heat Transfer in Body Heat Energy Harvesting

Natural convection heat transfer is only used when a low heat flux heat dissipation from the heat source is expected. Typical utilization is on body heat recovery but also applies to industrial low-grade waste heat recovery.

The poor heat transfer from the cold side surface is a significant bottleneck in heat transfer and thus power generation, especially when the thickness of the leg is limited due to the limited material flexibility, so that the design is hard to match the optimal TE thickness. The heat transfer coefficient of natural convection at the arms is discussed in this section based on the heat transfer correlation of several different geometries. Also, discussion on radiation heat transfer and thermal conduction follows in a special case.

3.5.1 *Natural air convection from human body*

In general, the majority of human body surface is vertically oriented. The natural air convection heat transfer coefficient of the vertically oriented plate with the other side insulated was given by Churchill and Usagi [23] in 1972. The correlation is based on the characteristic length (L_w). The average heat transfer rate decreases with decreasing surface area because the buoyancy force decreases due to the reduced air pressure difference between the top and the bottom (Fig. 3.14).

The Grashof number and the local Nusselt number at position x are given, respectively, by

$$\mathrm{Gr} = \frac{g\alpha\,(T_w - T_a)\,L_w^3}{\nu^2} \tag{3.45}$$

$$\mathrm{Nu}_x = 0.503\,(\mathrm{GrPr})^{1/4}\left[1 + (0.492/\mathrm{Pr})^{9/16}\right]^{4/9} \tag{3.46}$$

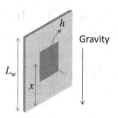

Fig. 3.14:   Natural air convection with a vertically oriented plate.

where g is the acceleration due to gravity, α is the coefficient of expansion, ν is the kinematic viscosity, L_w is the vertical length of the plate, T_w and T_a are, respectively, the wall and ambient temperatures, and Pr is the Prandtl number.

Then, the local heat transfer coefficient is obtained from the local Nusselt number as

$$h_x = \frac{\beta_f}{x} \mathrm{Nu}_x \tag{3.47}$$

The average heat transfer coefficient at the area of interest is found as

$$h_m = \frac{1}{L_2 - L_1} \int_{L_{12}}^{L} h_x dx \tag{3.48}$$

For example, the mean heat transfer coefficient of a vertical $L_w = 60$ cm high wall is 4.57 W/(m^2 K) for a skin temperature of 35°C and air temperature of 24°C. This correlation is alternatively used for a vertically oriented cylinder surface.

Another characteristic of heat transfer from human arms is due to the horizontally oriented cylindrical geometry of the arm (Fig. 3.15). There are several references for the Nusselt number correlation, and we are using the most recent and simplified model [24] as

$$\mathrm{Gr}_D = \frac{g\alpha \left(T_w - T_a\right) D^3}{\nu^2} \tag{3.49}$$

$$\mathrm{Nu}_D = 0.945 \left(\mathrm{Gr}_D \mathrm{Pr}\right)^{0.168} \tag{3.50}$$

$$h = \frac{\beta_f}{D} \mathrm{Nu}_D \tag{3.51}$$

Fig. 3.15: Natural air convection with a horizontally oriented cylinder shape, representing the lower arm lifted.

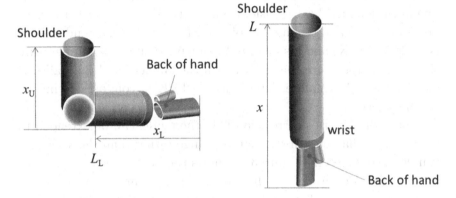

Fig. 3.16: Configuration of arms from the heat transfer point of view.

For example, with a diameter of 8 cm such as that of a wristband, the heat transfer coefficient is found to be 2.9 W/(m² K) for the same temperature conditions as above. Convective heat transfer from a human arm is modeled in variation of the configurations shown in Fig. 3.16.

3.5.2 *Pin-fin enhancement*

Pin-fin surface extension for passive cooling is a versatile method for any orientation of the surface relative to the gravity direction. The pin fin with a circular cross-section is investigated as it is the most robust for orientation. According to Ref. [25], the Nusselt number Nu_p for pin-fin arrays with a circular cross-section is found by using the modified Rayleigh number Ra* defined as

$$\text{Ra}^* = \frac{\text{Pr}g\beta\left(T_w - T_a\right)s_h^4}{L\nu^2} \qquad (3.52)$$

$$\left(\frac{\pi d}{2s_v}\right) \mathrm{Nu}_p = \frac{\mathrm{Ra}^{*3/4}}{20} \left[1 - e^{-\frac{120}{\mathrm{Ra}^*}}\right]^{1/2} + \frac{\mathrm{Ra}^{*1/4}}{200} \qquad (3.53)$$

$$h_p = \mathrm{Nu}_p \frac{\lambda_f}{s_h} \qquad (3.54)$$

where s_h is the center-to-center pin-fin spacing.

The total heat amount dissipated from the pin-fin array is found by multiplying the heat transfer coefficient h_p by the exposed surface area of each fin and the number of pins. The air thermal conductivity is 0.026 W/(m K) and assumed to be independent of the temperature since the temperature change between the ambient and the wall surface is relatively small, e.g. ranging 10−15 K in comfortable indoor circumstances.

Based on Aihara's correlation [26], the model can be extended to a very small diameter in the 100 μm range rather than the scale of a couple of millimeters. Figure 3.17 shows the heat transfer coefficient for a highly dense and small diameter pin-fin arrays in a variety of lengths. The shaded area shows the out of valid range, i.e. Ra* number is smaller than 0.3. At small diameters, the apparent heat transfer coefficient increases rapidly with decreasing diameter. Since the natural convection in such small diameter scales has not been

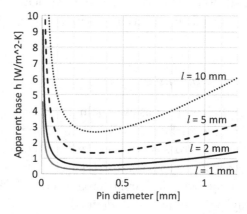

Fig. 3.17: An apparent base plate heat transfer coefficient vs pin-fin diameter in a variety of pin lengths. The region smaller pin diameter less than 0.8 mm is the out of valid range that Aihara *et al.* indicated.

explored yet, pin fins like human hairs may be an interesting scale
to investigate as a future work.

Considering an organic material with thermal conductivity
greater than 0.2 W/(m K), if the pin-fin diameter $d = 1$ mm and the
spacing, s_h and s_v, in horizontal and vertical orientations, respectively, are 1.6 mm each, the apparent heat transfer coefficient is
2.6 W/(m^2 K) for length 5 mm and 5.0 W/(m^2 K) for length 10 mm.
The pin fins are required to be 1 cm long to properly dissipate the
static mode heat flux ~ 40 W/m^2. This length is somewhat inconvenient for the thin power generator but could be possible with an
elegant design of the heat sinks.

3.5.3 *Add-on radiation and conduction*

Radiation heat transfer can make a significant contribution to the
net heat transfer in a TE system when the surface is under passive natural convection and the surface has high emissivity (0.8–1.0)
(Fig 3.18). This is particularly true at high surface temperatures but
also can happen at moderate temperatures even for temperature differences of 10–20°C. Radiation can be significant from the human
skin as well [27] because it has a very large emissivity, which is 0.99
or more according to the recent study [28] in the infrared wavelengths
at body temperature.

Radiation heat exchange can also occur between skin and a wall
at the room temperature. The view factor Γ in this case may be
approximated to be $\varphi_{12} \sim 1$ by modeling the flat wall approximately
as a cylinder with a very large radius, e.g. $R_2 = 2$ m, while the human

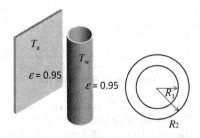

Fig. 3.18:   Radiation heat transfer from vertically oriented arm to a wall nearby.

arm is modeled as another concentric cylinder with a smaller radius, e.g. $R_1 = 0.25$ m, as shown in Fig. 3.18. The emissivities of a painted drywall or a concrete wall usually range 0.9−0.98. Clothes such as shirts made of cotton also show a similar emissivity of ~ 0.95. In this study, the emissivity of body is intentionally set to a relatively conservative value 0.95.

3.5.4   *Effective heat transfer coefficient*

The effective heat transfer coefficient should include natural/forced convection and radiation heat transfer and these are added into one coefficient (Fig. 3.19). Three locations such as wrist, elbow, and upper arm, and five situations such as standing up, sitting on a chair, working on a desk, walking with slightly swinging arms (0.6 m/s at the wrist), and running at 11 m/s speed are the cases commonly investigated. The application near the elbow may have a leading edge for the natural convection and it is a promising location for higher heat transfer across the human arm. The summary of the calculations is shown in Table 3.1. For the upper arm, a 1/4 of the cylinder surface is considered adiabatic while the rest of the surface area is exposed to convection. The size of TEG module may not always match the area of heat harvesting, a lateral thermal spreader may be necessary

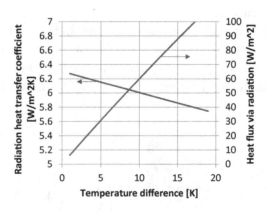

Fig. 3.19: Effective heat transfer coefficient by radiation heat transfer at a body skin temperature of 35°C and air temperature of 24°C. The heat flux caused by the radiation thermal cross-talk to the wall causes a very similar range of natural convection at a temperature difference of around 10°C.

Table 3.1:   Calculated effective heat transfer coefficients [W/(m² K)] for wearable TE energy harvesters at several different locations shown in Fig. 3.20 and wearer's activities.

Case	Wristband 25 cm × 2 cm	Elbow cover 12 cm × 12 cm	Upper-arm band 12 cm × 18 cm
Stand	7.99	9.36	8.91
Laptop	8.08	9.36	8.91
Desk	27.99	9.36	8.91
Walk (0.6 m/s)	30.81	13.34	9.43
Running (11 m/s)	98.93	98.93	98.93

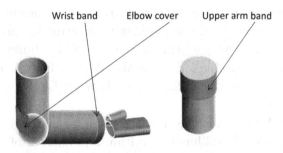

Fig. 3.20:   Locations of wearable thermoelectric energy harvesters used for the effective heat transfer coefficient calculations summarized in Table 3.1.

to fully obtain the power generation potential. The thermal spreader could be made of a layer of high thermal conductive material, e.g. graphite sheet.

3.5.5   *Convective heat transfer enhancement*

Based on the above modeling and analysis, approaches are further discussed for the enhancement of the effective heat transfer coefficient of a wearable thermoelectric energy harvester at different locations in the following sections.

3.5.5.1   *Wrist*

In the earlier approach, a watch-head-type wrist body heat harvester was investigated by the MIT group [29] and Leonov *et al.* [30], which used only a fractional area of the wrist for generating power. The wrist has a similar temperature around it and can be considered as a cylindrical uniform temperature heat source. The active area, in fact, can be much larger than the size of the watch head and could reach $\sim (24 \times W)$ cm^2 if the entire watch strap is developed as a harvester. This may require development of flexible thermoelectric materials for the harvester to conformally attach on the curved human skin [31]. The power output is proportional to the contact area. For example, a recent healthcare wristband equipment [32] has a 50 cm^2 contact area for the band.

Heat fins are a conventional technique to extend the heat transport surface for enhanced heat transfer coefficient in natural convection, but they will not be linearly effective for enhancing the radiation. Despite the performance enhancement, bulky and heavy fins may not be desirable for a wearable device. The key design concerns for the fins are, therefore, the minimum height required and the orientation.

A thermal boundary layer [33] exists for natural convection, within the range of a millimeter. Within a millimeter of height from the surface, the heat fins may not work effectively. The fins have to allow airflow through the fin gaps. Hence, it is desirable to have the fins aligned with the gravity force for natural convection. In the case of a wristband-type device, the relative orientation of fins can change much by the wearer's activities, so pin fins [34] might be a better solution for mixed vertical and horizontal orientations.

3.5.5.2   *Arm and elbow*

The upper arm has a relatively large area and obviously could provide a larger power output. For a fin surface design, the location can be considered as almost always a vertical wall. Recent shoulder-mount gear allows one to put smartphone on the upper arm. TE generator could be integrated with such device but not sufficiently supply power to charge a smart phone. Rather, the TE generator on the upper arm may be suitable to operate some monitoring sensors and wireless communication devices.

The largest heat transfer coefficient may be found at the elbow especially at the location indicated in Fig. 3.20. It may be an interesting area to investigate in heat transfer standpoint. However, at the same time, the skin of elbow stretches a lot. The mechanical solution to match the body parts will be a big challenge.

3.6 Summary

Efficient heat transfer is a key contributor to the performance of thermoelectric modules and systems. The performance of attached heat sinks is particularly important because they essentially determine the terminal temperature of the thermoelectric legs. Internal heat conduction drives the temperature gradient across the elements, which determines the power output and the material volume used, thus the material cost of the module. Spreading resistances and the convective heat transfer at the hot and cold sides are the factors that limit the heat flow in the system. In addition, electrical parasitic losses are directly related to the design of metal electrodes as well as TE legs. Passive heat transfer with natural air convection is modeled for human body heat recovery based on the geometry and the human activity conditions. This analysis on the heat transfer is in principle applicable for other low-grade heat recovery systems, where the difference is the orders of magnitude of the heat flux.

References

[1] J. P. Hartnett, W. M. Rohsenow, E. N. Ganic and Y. I. Cho, *Handbook of Heat Transfer*, New York, NY: McGraw-Hill (1973).
[2] Yazawa, Kazuaki and Ali Shakouri, Cost-effective waste heat recovery using thermoelectric systems, *J. Mater. Res.*, **27**(9), 1211–1217, (2012).
[3] T. J. Hendricks, New paradigms in defining optimum cost conditions in thermoelectric energy recovery designs, *Mater. Today: Proc.*, **8**, 613–624 (2019).
[4] Y. S. Muzychka, M. M. Yovanovich and J. R. Culham, Thermal spreading resistance in compound and orthotropic systems, *J. Thermophys. Heat Trans.*, **18**(1), 45–51 (2004).
[5] S. Song, S. Lee and V. Au, Closed-form equation for thermal constriction/spreading resistances with variable resistance boundary condition, *Proceedings of the 1994 International Electronics Packaging Conference*, Atlanta, Georgia, 111–121 (1994).

[6] M. H. N. Naraghi and V. W. Antonetti, Macro-constriction resistance of distributed contact contour areas in a vacuum environment, *ASME-Publications-HTD*, **263**, 107–107 (1993).

[7] D. B. Tuckerman and R. F. W. Pease, High-performance heat sinking for VLSI, *IEEE Electron Device Lett.*, **2**(5), 126–129 (1981).

[8] J. Lee and I. Mudawar, Assessment of the effectiveness of nanofluids for single-phase and two-phase heat transfer in micro-channels, *Int. J. Heat Mass. Tran.* **50**(3–4), 452–463 (2007).

[9] K. Yazawa and A. Shakouri, Cost-efficiency trade-off and the design of thermoelectric power generators. *Environ. Sci. Technol.*, **45**(17), 7548–7553 (2011).

[10] W. M. Kays and A. L. London, *Compact Heat Exchangers*, Third Edition (1984).

[11] R. K. Shah and A. L. London, Laminar Flow Convection in Ducts, *Advances in Heat Transfer, Supplement* 1, T. F. Irvine and J. P. Harnett, Eds., Academic Press, San Diego, CA (1978).

[12] E. N. Sieder and G. E. Tate, Heat transfer and pressure drop of liquids in tubes, *Ind. Eng. Chem. Res.*, **28**(12), 1429–1435 (1936).

[13] Incropera, DeWitt, Bergman, Lavine, Internal flow [Chapter 8], in *Fundamental of Heat and Transfer*, 6th edition, John Wiley and Sons, pp. 512–513 (2007).

[14] L. E. Bell, Use of thermal isolation to improve thermoelectric system operating efficiency. In Twenty-First International Conference on Thermoelectrics, 2002. *Proceedings ICT'02, IEEE*, 477–487 (2002).

[15] G. Brown, The History of the Darcy−Weisbach Equation for Pipe Flow Resistance, *Proc. Environ. Water Res. Hist.*, 38 (2002).

[16] M. Barry, J. Ying, M. J. Durka, C. E. Clifford, B. V. K. Reddy and M. K. Chyu, Numerical solution of radiation view factors within a thermoelectric device, *Energy*, **102**, 427–435 (2016).

[17] F. Meng, L. Chen and F. Sun, A numerical model and comparative investigation of a thermoelectric generator with multi-irreversibilities, *Energy*, **36**(5), 3513–3522 (2011).

[18] J. A. Potkay and R. D. Sacks, *J. Microelectromech. Sys.*, **16**(5) (2007).

[19] T. Caillat, J.-P. Fleurial and A. Borshchevsky, Preparation and thermoelectric properties of semiconducting Zn4Sb3. *J. Phys. Chem. Solids*, **58**(7), 1119–1125 (1997).

[20] J. Sakamoto, G. Snyder, T. Calliat, J. eP. Fleurial, S. Jones and J.-A. Palk, System and method for fabrication of high-efficiency durable thermoelectric devices. US Patent Application, US 20060157101A1 (2006).

[21] H. H. Berger, Models for contacts to planar devices, *Solid State Electron. Lett.*, **15**(2), 145–158 (1972).

[22] M. Ono, A. Nishiyama and A. Toriumi, A simple approach to understanding measurement errors in the cross-bridge Kelvin resistor and a new pattern for measurements of specific contact resistivity. *Solid State Electron. Lett.*, **46**(9), 1325–1331 (2002).

[23] S. W. Churchill and R. Usagi, A general expression for the correlation of rates of transfer and other phenomena, *AIChE*, **18**(6), 1121–1128, 1972.

[24] S. O. Atayılmaz and I. Teke, Experimental and numerical study of the natural convection from a heated horizontal cylinder, *Int. Comm. in Heat and Mass Trans.*, **36**(7), 731–738 (2009).

[25] S. W. Churchill and R. Usagi, A general expression for the correlation of rates of transfer and other phenomena, *AIChE*, **18**(6), 1121–1128 (1972).

[26] T. Aihara, S. Maruyama and S. Kobayakawa, Free convective/radiative heat transfer from pin-fin arrays with a vertical base plate, *Int. J. Heat Mass Trans.*, **33**(6), 1223–1232 (1990).

[27] D. Mitchell, C. H. Wyndham, T. Hodgson and F. R. N. Nabarro, Measurement of the total normal emissivity of skin without the need for measuring skin temperature, *Phys Med. Biol.*, **12**(3), 359 (1967).

[28] F. J. Sanchez-Marin, S. Calixto-Carrera and C. Villaseñor-Mora, Novel approach to assess the emissivity of the human skin, *J. Biomed. Opt.*, **14**(2), 024006 (2009).

[29] Available at: http://www.gizmag.com/wristify-thermoelectric-bracelet/295 43/, as of March 28 (2015).

[30] V. Leonov, P. Fiorini, S. Sedky, T. Torfs and C. Van Hoof, Thermoelectric MEMS generators as a power supply for a body area network, *Transducers'05*, **1**, 291–294 (2005).

[31] J-H. Bahk, H. Fang, K. Yazawa and A. Shakouri, Flexible thermoelectric materials and device optimization for wearable energy harvesting, *J. Mater. Chem. C*, **3**(40), 10362–10374 (2015).

[32] Available: https://www.fitbit.com/surge, as of 26th, March (2015).

[33] W. M. Kays and M. E. Crawford, Chapter 17 Free-Convection Boundary Layers, *Convective Heat and Mass Transfer*, 3rd ed., McGraw Hill, 396–416 (1993).

[34] R. Bahadur and A. Bar-Cohen, Thermal design and optimization of natural convection polymer pin fin heat sinks, *IEEE Trans. Compon. Pack. Technol.*, **28**(2), 238–246 (2005).

Chapter 4

Refrigeration

This chapter discusses the modeling and analysis of thermoelectric refrigeration primarily for electronics cooling. Refrigeration in a thermoelectric device is driven by applying an electrical bias to the device to cause an active heat pumping. The heat pumping draws heat from the target device to be cooled and releases the heat to a thermal ground, e.g. to ambient air. The heat source temperature may be higher or lower than that of the thermal ground as the active cooling can transfer heat against the passive heat conduction. The model is developed for a single thermoelectric element (leg), which is then extended to a module and a system. The working principle is discussed, and then the methodology for design optimization of thermoelectric coolers is provided. The coefficient of performance (COP) is used to evaluate the performance since it is a common metric for conventional vapor-compression-based refrigeration. There are two objectives, COP and maximum cooling capacity Q_{max}, for design optimization. Then, a unique beta-shape curve in the relationship between COP and Q_{max} is discussed. Finally, the advantages of thermoelectric coolers in comparison with vapor compression cycles are discussed.

4.1 Working Principle

Thermoelectric refrigerators or coolers are sometimes called 'Peltier coolers' since it is based on the physical phenomenon called the Peltier effect named after the physicist, Jean Charles Athanase Peltier, who discovered it in 1834 [1]. When an electric current flows through a junction between two dissimilar materials, heat is either dissipated (Peltier heating) or absorbed (Peltier cooling) depending on the current direction. In cooling, the absorbed heat is transferred and rejected to the other side of the system through a heat sink and then to a thermal ground. Thermoelectric cooling modules have structures similar to thermoelectric power generators, typically consisting of multiple thermoelectric elements (legs) connected with metal electrodes, sandwiched by thermally conductive substrates at both sides. The modules typically have two flat and rigid substrates, which allow good thermal contacts to both the target and the heat sink. In a module, pairs of legs made of n-type and p-type thermoelectric materials are placed between the two substrates thermally in parallel and electrically in series with the patterned electrodes to form the circuit. A single pair configuration of legs is shown in Fig. 4.1. The heat Q is absorbed at the top contact and heat Q' is rejected at the bottom contact in this figure, where the temperature of the top contact, the target temperature, is one of the design objectives.

The legs form a closed electrical circuit with the electrodes on top and bottom (see Fig. 4.1), connecting an external power supply. Inside the legs, charge carriers are forced by the external bias to flow along the legs. The charged carriers are the excess free electrons in an n-type semiconductor or the excess holes (voids of electrons), which are positively charged, in a p-type semiconductor. Electrons move from top to bottom in the n-type element shown in Fig. 4.1, and holes

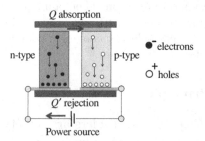

Fig. 4.1:   Schematic diagram of n-type and p-type thermoelectric pair legs. The external power source continuously causes the charge carriers, electrons in n-type and holes in p-type, to flow from top to bottom in both legs, which carry heat in the same direction to eventually cool the top side.

move in the same orientation, also from top to bottom in the p-type element in the figure. As the carriers move, they carry heat with them. Hence, the heat is carried from top to bottom. In view of electric current flow, electrons move in the opposite direction and holes move in the same direction of the current. Since the power source continuously supplies the power, the heat is continuously pumped from top to bottom in both n-type and p-type elements. As a result, the top side is cooled, and the bottom side is heated. Typically, a heat sink is attached at the bottom side for efficient heat removal from there.

Doped semiconductors are used for thermoelectric materials since the doping level is tunable in a wide range. Hence, the thermoelectrical properties vary quite different from the intrinsic properties of the pure crystals depending on the doping process and the impurity.

4.2 Design Optimization for COP and $Q_{\max}$

The performance of a thermoelectric material is evaluated with the dimensionless figure of merit known as the ZT value defined as

$$ZT = \frac{\sigma S^2}{\beta} T \tag{4.1}$$

where S is the Seebeck coefficient, σ is the electrical conductivity, β is the thermal conductivity, and T is the absolute mean temperature across the thermoelectric (TE) elements. The numerator σS^2 is referred to as the thermoelectric 'power factor'. In refrigeration or heat pump applications, the temperature dependency of the material properties is usually negligible since the temperature gradient is small and the properties vary with temperature only slowly. However, it is important to consider the temperature dependency if the temperature gradient across the thermoelectric leg is large:

$$Z(T)T = \frac{\sigma(T)S(T)^2}{\beta(T)} T \tag{4.2}$$

The figure of merit ZT of the material used largely determines the cooling performance. However, ZT is not the only factor that determines the refrigeration performance. The energy conversion process involves various impacts of the thermal interfaces with the hot and cold side heat reservoirs. For a heat pumping from lower to higher temperature, the thickness d of the legs along the heat flow direction is a key design parameter. If the thickness is too thin ($d \to 0$), the thermal resistance is significantly small and therefore the pumped heat can easily flow back due to the reverse temperature gradient. If the thickness is too thick ($d \to \infty$), thermal insulation is good, but the electrical resistance can be too large, which would create too much Joule heating in the legs and consume too much power and thus eventually exhibit a very low efficiency (small heat pumping with a large power input). Hence, there is an optimal design to maximize the heat pump performance. There are two different objectives: one is maximizing the cooling capacity $Q_{\max}$, and the other is maximizing the coefficient of performance (COP), the ratio of pumped heat (cooling capacity or cooling power) to the electrical power input.

4.2.1   *Analytical model*

The analytical model for a single-leg TE cooler consists of a thermoelectric element (leg) with thickness d, which is placed between the hot and cold reservoirs. At the starting point of the operation conditions, both the heat source and the cold reservoir temperatures are considered to be given and constant. However, for electronics cooling, the heat source temperature is designed to be equal to or lower than the maximum allowable temperature $T_{j,\max}$. In this case, the maximum possible heat removal needs to be determined for the given maximum allowable temperature. For room-temperature cooling, the cooled side temperature T_s is similar to the ambient temperature T_a ($T_s \approx T_a$) or even lower than the cold reservoir ($T_s < T_a$).

In order to investigate the quantitative performance estimation, an electrothermal combined network model can be used. Figure 4.2 shows the network with the temperature profile. At the two terminals of the leg, T_h is the temperature at the heat source (cooling target) and T_c is the one at the opposite side heat sink. Once the cooling operation is done properly at steady state, T_h must be lower than T_s in order to pump heat from the target, and T_c must be higher than T_a also to allow the heat to dissipate to the ambient. In the case

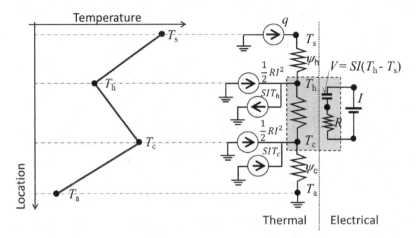

Fig. 4.2: Electrothermal resistance network model for thermoelectric refrigeration with the temperature profile in the heat flow direction and various heat flows involved in the network. This figure shows the case of an n-type thermoelectric leg.

of a room-temperature refrigeration, T_s can be lower than T_a, while a heat pumping for electronics cooling as another example may be designed for the case of $T_s > T_a$. The temperature difference between T_h and T_c is controlled by the TE module with an externally applied electrical power.

There is Joule heating that occurs uniformly inside the leg due to the electrical current flow in addition to the conduction heat flow due to the temperature difference, $(T_c - T_h)$. The Joule heating in total of RI^2 is equally dissipated to both sides, so a half Joule heat $0.5\ RI^2$ comes out to each side. The Peltier heat SIT is applied at each node with the direction depending on the direction of the current flow. In refrigeration, heat removal at node T_h and heating at node T_c happens by the Peltier effect.

Variation of the TE boundary temperatures T_h and T_c as a function of drive current is shown in Fig. 4.3. Figure 4.3(a) shows an example case of thermoelectric refrigeration, where the cooling target temperature is lower than ambient $(T_s < T_a)$. Figure 4.3(b) shows an example of heat pumping, such as electronics cooling when $T_s > T_a$.

Heat balance equations are formulated using the thermal circuit model at each of the four nodes at T_s, T_h, T_c, and T_a [K] in

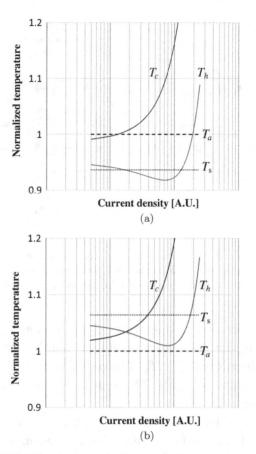

Fig. 4.3: Internal temperature profile (T_h and T_c) at the two sides of a single-leg TE cooler normalized by ambient temperature (T_a) as a function of drive current I. The ambient temperature is fixed at 40°C (313 K) and the heat source temperature is fixed at 20°C (293 K) in (a) and at 60°C (333 K) in (b). When T_h exceeds T_s, there is no longer net refrigeration.

Eqs. (4.3)–(4.6). Here, the external thermal resistances ψ_h and ψ_c [(K m^2)/W] at the hot and cold sides, respectively, are used. All these equations are based on per unit cross-section area:

$$Q_h = \frac{(T_s - T_h)}{\psi_h} \tag{4.3}$$

$$Q_h - SIT_h + I^2 R/2 = K(T_h - T_c) \tag{4.4}$$

$$K(T_h - T_c) + SIT_c + I^2 R/2 = Q_c \qquad (4.5)$$

$$Q_c = \frac{(T_c - T_a)}{\psi_c} \qquad (4.6)$$

with

$$K = \frac{\beta}{d} \qquad (4.7)$$

$$R = \frac{d}{\sigma} \qquad (4.8)$$

where S is the Seebeck coefficient [V/K] and I is the electrical current [A] applied from the external supply. Q_h is the heat removed from the heat source at T_s [K] and Q_c [W] is the heat rejected through the heat sink to T_a [K]. K is the thermal conductance per unit area [W/(m^2 K)] of the TE leg, β is the thermal conductivity [W/(m K)], R is the internal electrical resistance for a unit area [Ω m^2] of the leg, σ is electrical conductivity [1/(Ω m)], and d is the leg thickness [m].

The terms in Eqs. (4.4) and (4.5) involved with K follow the Fourier law by the temperature gradients. By the Peltier–Seebeck effect, heat energy is transported between T_h and T_c based on these temperatures and the applied current. The terms consisting of electrical resistance R represent the Joule heat occurring inside the volume of the leg.

The external bias that is required to let the current I flow in the leg must overcome the internal Seebeck voltage developed due to the temperature difference $(T_h - T_c)$ across the leg, which is $V_{\text{Seebeck}} = -S(T_h - T_c)$. Here, a minus sign has been applied as we assume an n-type leg in the model with Seebeck coefficient to be negative ($S < 0$). Thus, the total voltage supply becomes $IR - S(T_h - T_c)$. Then the total power consumption for the given current I becomes

$$W = I(IR - S(T_h - T_c)) = I^2 R - SI(T_h - T_c) \qquad (4.9)$$

The heat input Q_h is eventually removed from the heat source, while the applied power is W. Hence, the cooling coefficient of performance

COP is given by

$$\text{COP} \equiv \frac{Q_h}{W} = \frac{Q_h}{I^2 R - SI(T_h - T_c)} \tag{4.10}$$

From the previously shown energy conservation, by substituting Eq. (4.3) into Eq. (4.4),

$$T_h = \frac{-2T_a + 2((K - SI)\psi_c + 1)T_c - R\psi_c I^2}{2\psi_c K} \tag{4.11}$$

Similarly, substituting Eq. (4.6) into Eq. (4.5),

$$T_c = \frac{-2T_s + 2((K + SI)\psi_h + 1)T_h - R\psi_h I^2}{2\psi_h K} \tag{4.12}$$

These equations still recursively referred to each other. Further solving the equations with respect to T_h by substituting Eq. (4.12) into Eq. (4.11),

$$T_h = \frac{\psi_c\psi_h R S I^3 - (2\psi_c K + 1)\psi_h R I^2 + 2\psi_c S T_s I - 2((\psi_c T_s + \psi_h T_a)K + T_s)}{2(\psi_c\psi_h S^2 I^2 + (\psi_c - \psi_h)SI - (\psi_c + \psi_h)K - 1)} \tag{4.13}$$

As above, the hot side temperature T_h is found as a cubic order function of the applied electrical current I with given boundary T_s and T_a.

At the limit of $I \to 0$ and $d \to 0$ (hence $K \to \infty$), Eq. (4.13) converges to

$$T_h \to \frac{\psi_c T_s + \psi_h T_a}{(\psi_c + \psi_h)} \tag{4.14}$$

and $T_h \to T_c$. This temperature is the mean temperature of the nodes determined by the two external thermal resistances.

The heat amount removed/absorbed from the heat source Q_h is found by using Eq. (4.3) as T_h is found by using Eq. (4.11). Likewise, the required heat rejection from the refrigeration system Q_c is determined by using Eq. (4.6).

Then, T_c as a function of T_s and T_a is found by substituting Eq. (4.13) into Eq. (4.12):

$$T_c = \frac{\psi_c \psi_h R S I^3 - (2\psi_h K + 1)\psi_c R I^2 - 2\psi_h S T_a I - 2((\psi_c T_s + \psi_h T_a)K + T_a)}{2(\psi_c \psi_h S^2 I^2 + (\psi_c - \psi_h)SI - (\psi_c + \psi_h)K - 1)}$$

$$(4.15)$$

Equations (4.13) and (4.15) for T_h and T_c are only the functions of given variables, $T_s, T_a, \psi_h, \psi_c, K, R, S$, and I. The heat removal from source side (cooling target) Q_h is determined by substituting Eq. (4.14) into Eq. (4.3). Also, the heat rejection on the other side q_c is determined by substituting Eq. (4.15) into Eq. (4.6).

Until here, all the equations are based on a unit area. In practice, however, the real dimensions really matter for estimating the scale of handling heat amount. Also, here, the fill factor F is introduced, which is the fractional area coverage of thermoelectric legs over the footprint area A [m^2]. With the number of legs N in a module, total heat flux removed from the heat source q_h [W/m^2] is found as

$$q_h = NFQ_h/A \qquad (4.16)$$

Similarly, heat flux rejected through heat sink to the thermal ground is

$$q_c = NFQ_c/A \qquad (4.17)$$

In this formulation, p-type and n-type legs are assumed to have the same material properties and the only difference is the sign of Seebeck coefficient (positive or negative). In reality, the material properties are most likely different. Then the p-type and n-type elements must be separately solved for, and then the lateral heat transfer between adjacent p-type and n-type elements that occurs due to the difference in heat inputs in each type must be taken into account. One model with the average values of the p-type and n-type properties can still be useful for a reasonably good performance prediction if they are made of the same kind of semiconductor materials.

Key electrical properties of the module with subscript m are the total internal resistance, the current, the voltage supply and the total power supply given by the following:

$$R_m = R(2N/A_e) \tag{4.18}$$

$$I_m = I A_e \tag{4.19}$$

$$V_m = I_m R_m - S I_m (T_h - T_c) \tag{4.20}$$

$$W = V_m I_m \tag{4.21}$$

where A_e is the cross-sectional area of one leg ($= FA$). The unit of all the parameters with subscript m is no longer per unit area. Here, any parasitic losses and the impact of substrates are excluded.

An interfacial spreading thermal resistance may also need to be considered as a part of the external resistance for the heat sink. The hotspot area, by definition, is smaller than the footprint area of heat sink. Our focus is on the COP of TE hotspot cooling. Since the impact of heat dissipation from the TE cooler is much less important than the heat sink's heat transfer performance, this factor can be ignored for design optimization. The design of the heat sink must be matched to the cooling target. A study of the heat sink design and its impacts on hotspot cooling was reported in Ref. [2]. The use of heat spreading is relatively more important for the integration of thermoelectric in electronics packaging. Detailed investigation of lateral spreading and electron transport was summarized in Ref. [3].

To capture the characteristics of thermoelectric refrigeration, an example of refrigeration design with a typical thermoelectric module is discussed here. As an example, thermal resistance at the contact with the heat source (refrigeration target) is 1.0×10^{-4} (Km^2)/W or 1.0 (K cm^2)/W and that of the other side (heat rejection) is 2.0×10^{-4} (K m^2)/W. The size of heat source, thermoelectric module footprint, and the heat sink footprint are assumed identical for the simplicity in dimensions of 40 mm × 40 mm. The thermoelectric module contains 127 pairs of identical p-type and n-type legs (16 × 16 array) with leg length (height) is 3.0 mm. A vacant space for one pair is used for lead wires to connect to the external power supply. Fill factor of the legs is 33%. Thermoelectric properties used are 2.0 W/(m K) thermal conductivity, 1×10^5 S/m electrical conductivity, and 2.4×10^{-4} V/K Seebeck coefficient. The values are similar to those of bismuth telluride (Bi$_2$Te$_3$) in the market with well-optimized doping. The ZT

value at temperature of 20°C is 0.84. This value may sound too modest compared to the highest ZT values reported for recent nanostructured thermoelectric materials, but a consideration is made to include a realistic value that reflects the average performance of the commercially available materials. The target temperature varies from −50 K to +50 K relative to the cold reservoir temperature set at 300 K. For example, when cooling an electronic chip such as CMOS, the maximum allowable heat source temperature $T_{j-\max}$ is around 80–100°C [4, 5], above which the performance and durability of the chip is significantly reduced. In this case, the design objective is to remove heat at the junction as much as required for keeping the heat source temperature below $T_{j-\max}$ in the worst case. In contrast, for refrigeration of a target at a much lower temperature than room temperature (heat rejection reservoir), e.g. −30°C, the degree of cooling (ΔT) may be more important than the cooling power.

Figure 4.4 shows the temperature variation of the system at the four nodes for two cases of operation: the left-hand side plot shows the case when the drive current I is optimized for maximum heat removal $Q_{\max}$, and the right-hand side plot shows the case when I is

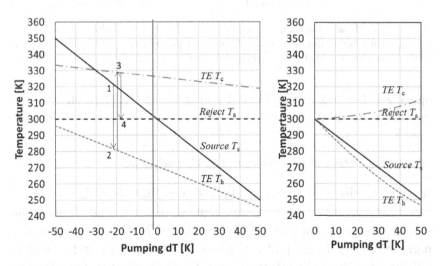

Fig. 4.4: Temperature profile of thermoelectric legs as a function of temperature difference. The sequence represents $1 : T_s \rightarrow 2 : T_h \rightarrow 3 : T_c \rightarrow 4 : T_a$. Left plot shows the case optimized for maximum heat removal and right plot shows the case optimized for COP.

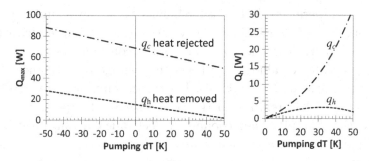

Fig. 4.5: Maximum heat removal as a function of temperature difference. Left plot shows the case optimized for maximum heat removal and right plot shows the case optimized for COP. Note that the scales are different.

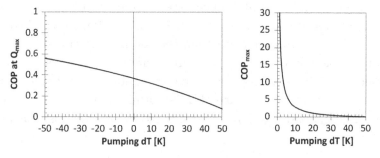

Fig. 4.6: COP as a function of temperature difference. Left plot shows the case optimized for maximum heat removal and right plot shows the case optimized for COP. Note that the scales are largely different.

optimized for maximizing COP. Figures 4.5 and 4.6 also show maximum heat removal and COP as a function of temperature difference.

4.2.2 *Relationship between COP and Q_{max}*

In utilizing the model into a design of refrigeration cycle or system, there are two major performance figures [6] to achieve. Among the system design parameters $\{T_s, T_a, \psi_h, \psi_c, K, R, S, I\}$, internal design parameters $\{K, R, S, I\}$ of a thermoelectric refrigerator (or cooler) fall into two key parameters d and I, assuming that the material properties are fixed. The thickness d directly changes both K and R, while I changes the rate of carrying heat.

Parametric analysis was carried out on the variation of thermal resistance ratio γ with changing applied current I. The thermal resistance ratio is defined as the ratio of the leg thermal resistance to the total external thermal resistance, such that

$$\gamma = (d/\beta)(\psi_h + \psi_c) \tag{4.22}$$

For clarification, the fill factor and the cross-section area of the leg are assumed to be fixed. Figure 4.7 shows COP vs Q_h by changing the driving current and γ. In this analysis, the material properties, i.e. the thermal conductivity, electrical conductivity, and Seebeck coefficient, are given, so that $ZT=1$. Since the ZT value includes the mean temperature multiplied [7], the ZT value slightly shifts as the driving current changes because the mean temperature changes. But the change is very small in the range shown in Fig. 4.7.

The plot in Fig. 4.7 only shows the first quadrant for positive Q_h and COP (net cooling) only. A smaller γ means a shorter TE leg. At a small I, the Q_h and COP are in fact negative (net heating) because the thermal conduction dominates over the Peltier and Joule

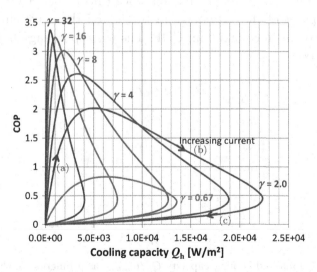

Fig. 4.7: COP vs Q_h (cooling capacity per unit area) with a valuation of γ by changing leg thickness d. $ZT = 1.0$, $\beta = 1.5$ [W/(m K)], 200 elements in a 40 mm × 40 mm module. Both heat exchanger performances are modeled as heat transfer coefficient 2000 [W/(m² K)].

heats while the heat pumping power is negligibly small with a very small current. With increasing I, the unique β-shaped curves are drawn, starting from a negative heat pumping negative COP, and after crossing $(0, 0)$, the COP keeps increasing steeply. There remains an opportunity to simultaneously enhance the cooling capacity Q_h and the COP by increasing the current. At a certain I, the curve hits the maximum COP and then the COP declines, but Q_h keeps further increasing. Then, the curve eventually reaches the maximum cooling capacity and turns toward the origin, $(0, 0)$. Finally, the curve passes over the origin point and to the third quadrant with the negative COP, and a large negative cooling capacity will be found if a very large current is applied. In Fig. 4.7, $Q_{\max}$ occurs at approximately $\gamma = 2.0$. This value is found under the condition with symmetric external thermal resistances, i.e. $\psi_h = \psi_c$. The optimum design condition varies with the ratio of the two external thermal resistances, as shown in Fig. 4.8.

In the design of a thermoelectric cooler for electronics cooling, the leg thickness d plays a significant role for the cooling performance. Figure 4.9 shows the impact of leg thickness on COP and Q_h. At each point of the 3D curves in the figure, the drive current has been optimized for maximum COP. The corresponding Q_h is shown in the y-axis.

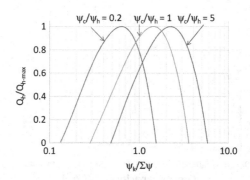

Fig. 4.8: Normalized cooling capacity $Q_h/Q_{h,\max}$ as a function of the thermal resistance ratio γ. The other conditions are the same as in Fig. 4.7 for the symmetric case. When the heat dump side heat sink is much better than the source side, the optimum cooling capacity is found at a smaller leg thickness. The other way around, the optimum is found at a larger leg thickness when the source side heat sink is better.

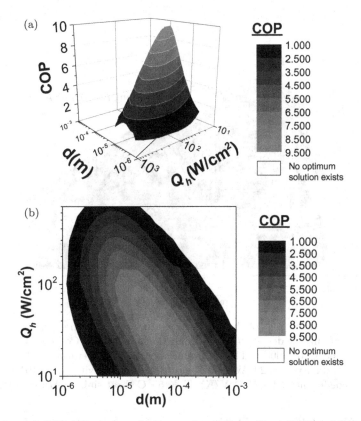

Fig. 4.9: (a) COP (3D plot) vs thickness, d, and Q_h at $T_h = 65°$C and (b) 2D top view of the same plot. Thermal conductivity is 1.5 W/(m K), electrical conductivity is 6.3×10^4 1/(Ω m), Seebeck coefficient is 2.8×10^{-4} V/K.

Typically, in electronics cooling, the heat source (cooling target) temperature must be maintained at or lower than the allowable maximum junction temperature T_{jc}. Two design parameters, which are the leg thickness d and the drive current I, give a unique curve under the thermal design conditions. Figure 4.10 shows the cooling side temperature T_h as a function of these parameters. There exists a minimum temperature on the three-dimensional curve. The minimum temperature point shall be sufficiently lower than the allowable maximum junction temperature T_{jc} (referred to as T_s in the analytic model) to give an allowance for the contact thermal resistance, which increases the junction temperature.

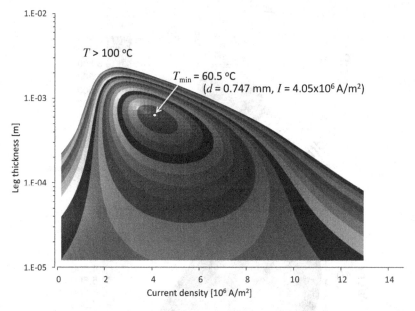

Fig. 4.10:   Contour plot of the chip temperature with thermoelectric cooling as functions of current density I [A/m^2] and thickness d [m]. Contour scale is 5°C step from 60°C to 100°C. The ZT value at target temperature is 1.02, where thermal conductivity is 2.0 W/(m K), electrical conductivity is 1×10^5 1/(Ω m), Seebeck coefficient: 2.4×10^{-4} V/K. $T_a = 55$°C (heat sink temperature).

4.3   Cost-Effective Design

The material mass used in a TE leg is proportional to the thickness d for the unit cross-sectional area. The material cost can be linked with the performance. Material cost Y_m [$] is given by

$$Y_m = \rho N d A Y_U \qquad (4.23)$$

where Y_U is the market raw material unit cost per mass [$/kg], NdA is the total volume of the material used, and ρ is the density of the material [kg/m^3]. The market price varies over time, depending on the abundance of the raw material and the process cost. For the applications near room temperature, bismuth telluride (Bi_2Te_3) and their alloys are typically used. The cost of these materials is approximately ~ 500 $/kg for a large volume module manufacturing. The density of Bi_2Te_3 is 7.74 g/cm^3 [8].

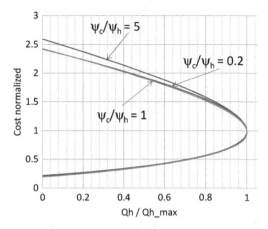

Fig. 4.11: Cost vs Q_{h_max}. The cost is normalized by the optimum cost for Q_{h_max} by changing thermoelectric element thickness. Lower cost represents thinner element. The dependency on external thermal resistance ratio ψ_c/ψ_h is very small.

4.3.1 *Impact of heat sink performance*

Figure 4.11 shows the material cost normalized by the optimum design for Q_{max} as a function of cooling capacity with varying leg thicknesses. If the relation between the cost and the COP is plotted, it will show a similar trend. Obviously, a thinner leg costs less but delivers lower performance as well. If the thickness is too thick, cost gets higher, and the performance Q_h reduces at the same time.

4.3.2 *Impact of figure of merit*

There are many approaches investigated to improve the figure of merit, e.g. researches including the reduction of thermal conductivity by introducing confined structures, such as nanowires [9], complex nanostructures [10], and nanowire heterostructures [11], or total improvement of the figure of merit [12]. In this section, the impact of the material property on cost−performance trade-off is investigated based on the analytical model. Figures 4.12 and 4.13 show the cost per unit performance Q_h and COP, respectively. Both include the cases $ZT \sim 1$ and $ZT \sim 2$, where the difference is only on the thermal conductivity.

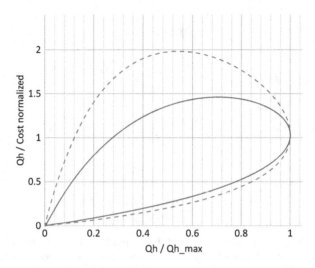

Fig. 4.12: Q_h (normalized)/Cost (normalized) vs Q_h/Q_{h_max}. The solid curve shows $ZT \sim 1$ and the broken curve shows the $ZT \sim 2$ with reduced thermal conductivity.

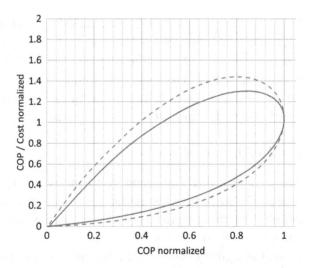

Fig. 4.13: COP (normalized)/Cost (normalized) vs COP/COP$_{\mathrm{max}}$. The solid curve shows $ZT \sim 1$ and the broken curve shows the $ZT \sim 2$ with reduced thermal conductivity.

4.4   Cooling Limit of Infinitely Large Z

Due to the moderate performance of thermoelectrics, improving the material is a natural logical thinking to obtain a better heat pump performance and efficiency (COP). This section discusses how far the change in material properties could make improvement, i.e. increasing the Z value. Here, we take an example case of pumping heat from a fluid flow with fixed temperature at 20°C to another fluid flow with fixed temperature at 40°C. Both effective heat transfer coefficients are assumed to be 3500 W/(m² K). In this simulation, thickness of the thermoelectric is given as constant to be 1.5 mm and the only design parameter other than ZT value is the drive current I. As shown in the beta-shape plot (COP vs Q_h) in Fig. 4.7, as the current increases, the COP hits the peak first. As the current keeps increasing, COP decreases, but Q_h still increases. Then, Q_h hits the peak at some point and both performances decrease afterwards. Figure 4.14 shows the variation of COP and Q_h with ZT value. Each contour represents a fixed ZT value with varying current. Starting from $ZT = 0.7$, the contour becomes larger with increasing ZT value. The most outer shape shows the case of $ZT \to$ infinity. COP is dependent on all design parameters, including the thermal resistance of heat sinks and the

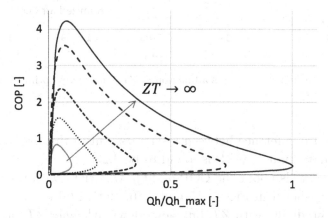

Fig. 4.14:   COP as a function of normalized heat removal Q_h. Note increasing ZT value with equal impact of three thermoelectric properties. Both converge at a value of infinitely large ZT. Curves represents from smaller to larger $ZT = 0.7$, 5.6, 45, 716, and ∞, respectively.

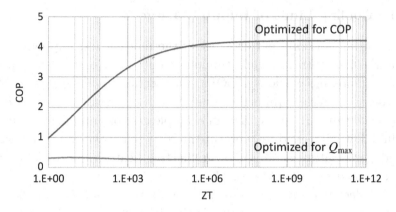

Fig. 4.15:   COP as a function of ZT. Curves are optimized for COP and Q_{max}, respectively.

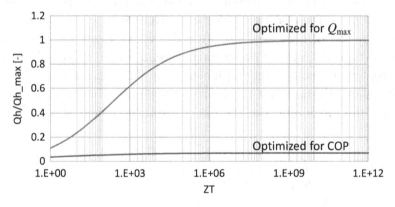

Fig. 4.16:   Normalized Q_{max} as a function of ZT. Curves are optimized for COP and Q_{max}, respectively.

temperature difference between reservoirs, but in this example case, the ultimate limit of COP is found to be 4.225.

More details about the COP and Q_h variations with varying ZT values are shown in Figs. 4.15 and 4.16. Both COP_{max} and $Q_{h,max}$ increase gradually with ZT, but slows down at higher ZT values and saturates. Improvement of COP by changing ZT from 0.7 to 5.6 is about 1.86 times. For the same ZT change, Q_{max} demonstrates a larger improvement, but still only 2.16 times.

4.5 Thin Films

Thin film thermoelectric devices are useful for microscale hotspot cooling. Key differences from common refrigeration are as follows: (1) the hot side temperature is most likely higher than the heat sink side, (2) the maximum allowable temperature is given to the hotspot, and (3) high heat density needs to be removed. An example of a stand-alone $500 \times 500 \ \mu m^2$ or $0.25 \ mm^2$ hotspot is investigated in the following analysis. We assume that the hotspot is maintained at $80°C$ by the cooling, while the heat sink temperature is $60°C$. The exact same size of a thermoelectric leg is attached to the hotspot with a thermal resistance, while the other side of the leg is in contact with the heat sink. Expecting each 10 K of the temperature difference across the contact interfaces, the effective external heat transfer coefficient is given as $5 \times 10^5 \ W/(m^2 \, K)$.

There is an optimum thickness for a high heat flux removal, and it is thinner than 10 μm in this case (Fig. 4.17). The thinner the leg is, the larger the current required, so that thin legs may pose a technological challenge as well. For example, a current of 85 A is required for this design point to remove 937 W from the 0.25 mm^2 hotspot. Removing 100 W from the same hotspot is possible with thickness around 0.3 mm and 5.6 A current. Cost-effective fabrication of the legs with thickness in the range of 10–100 μm is relatively

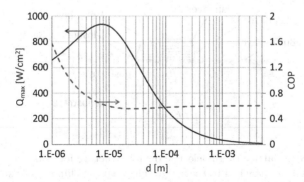

Fig. 4.17: Q_{max} and COP as functions of leg thickness. All data points of the curves are designed for maximum heat removal Q_{max}. The optimum thickness is found to be 7.56 μm for this example.

a challenge, while several fabrication techniques are available, such as sputtering, chemical vapor deposition, and so on. From a bulk solid, thinning a wafer down to tens of microns is technically possible, but may require precise alignment and delicate thinning techniques for uniform thickness and flat surfaces. The most popular thermoelectric material bismuth telluride is especially brittle and causes a big challenge in maintaining the mechanical stability of such a thin wafer. Development of non-brittle materials and a quick process to produce several hundreds of micron thick legs is highly desirable for hotspot cooling. Thin film deposition techniques such as sputtering [13] and chemical vapor deposition [14] would be the best option. Electrical contact resistances were not considered in this analysis, but, due to the high current density, Joule heating at the contact can be a big concern for practical utilization.

Thin films are also useful for modifying electron and phonon transport with multiple boundaries and interfaces to decouple the two transport regimes and enhance ZT [15]–[17]]. A superlattice thermoelectric device (see Fig. 4.18) was fabricated for hotspot cooling with microchannel heat removal from an entire chip [18]. The device consisted of 200 periods of $Si_{0.7}Ge_{0.3}/Si$ superlattices in 2.7 μm thickness. See Ref. [19] for the device details (Fig. 4.19).

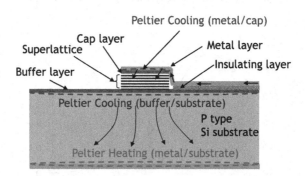

Fig. 4.18: Superlattice thermoelectric device structure for hotspot cooling. The top side contact is attached to the active side of silicon chip at the exact location and same size of hotspot. A microchannel heat sink in the substrate is expected to directly remove the heat from hotspot and the internally generated Joule heat.

Fig. 4.19: Optical image of fabricated superlattice coolers.

Source: Reproduced with permission from *Proc. IEEE*, **94**(8), 1613–1638 (2006).

4.6 Multi-stage Thermoelectric Coolers

A configuration of more than two modules stacked up is called multi-stage configuration [20]. Figure 4.20 shows an example of the appearance of multi-stage module.

Two effects are taken into account in multi-stage modules: (1) smaller temperature difference across each individual stack and (2) thermal spreading through stacking with geometrical spreading, i.e. larger area or a larger number of modules at a lower stage to reduce the heat flux (heat flow per unit area). As discussed in Section 4.2, the COP of heat pumping significantly changes due to changes in temperature difference across each stack. Thermoelectric heat pump works more energy efficiently, i.e. higher COP, when the degree of cooling (ΔT) is smaller. The total COP of a multi-stage system is obtained using the individual stage COP as

$$\text{COP}_{\text{system}} = \frac{1}{\displaystyle\prod_{i=1}^{N}\left(1 + \frac{1}{\text{COP}_i}\right) - 1} \tag{4.24}$$

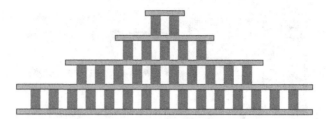

Fig. 4.20:   Example stack of multi-stage thermoelectric cooler module. Off-the-shelf example is found everywhere, such as Ref. [21].

Pumping a large heat flux requires a large drive current. The design of heat spreading helps to reduce the heat flux while total heat flow is maintained, which also helps to improve the system COP. Multi-stage thermoelectric coolers can therefore effectively improve the COP compared to a single-stage module. However, due to the contribution of interfacial resistances, the system performance can decrease as the number of stage increases [22], and finally, the pump heat flow can diminish. Hence, there must be an optimum number of stages to maximize the COP. The multi-stage, therefore, may not enhance the cooling power limit in heat flow, but may enhance the cooling temperature degrees by the multi-stage cooling without much increased input electrical power. A US patent shows a concept of deep cryogenic cooler with multi-stage thermoelectric devices [23].

4.7   Scalability for a Working Temperature Range

The universal principle of a TE cooler discussed above can be used for a wide range of temperatures. This is quite a unique characteristic of thermoelectric cooling, compared to any other cooling techniques, because they work only in a limited temperature range. Although there are various factors affecting the cooling performance, in principle, we can design a thermoelectric heat pump that can pump heat from an oil with temperatures of over 200°C or even electronics cooling under cryogenic conditions, e.g. $T < 77$ K, following the same principle and mechanism. This is impossible for vapor compression cycle because of the limited temperature window for the available refrigerants.

In practice, temperature-dependent material properties are also important for temperature scaling. The figure-of-merit ZT value itself is significantly temperature-dependent. Hence, the same material may not necessarily be usable for multiple different temperature ranges. Selecting a material suited for the specific operation temperature range is important to keep the performance high. The impact of temperature dependency of material properties has been investigated since 1959 [24]. The fundamental mechanism that causes the temperature dependency is explained in detail in Ref. [25].

Either in high- or low-temperature operations, it is typical that the system returns to near room temperature when the operation is turned off. In this case, thermal expansion of materials may cause local or even system-level dimension changes, which can potentially cause deformation of the cooler, and create a large stress at the mechanical contacts due to the coefficient of thermal expansion (CTE) mismatch. This is true for power generation as well. There has been a study about reducing the interfacial stress caused by thermal expansion mismatch in a TE system using a low fill factor design [26].

4.8 Transient Cooling

When the Peltier effect happens at a contact, the time response of changing temperature at the contact is relatively fast since the thermal mass involved at the contact is significantly smaller. In contrast, Joule heating occurs in the entire volume of the electric current flow path and thus the large thermal mass involved would respond relatively slowly in the beginning of a step bias input. The time to reach the equilibrium is ruled by the thermal diffusion. Thus, it is important to know how much thermal mass is involved along the heat flow between thermoelectric cooling contact and the cooling target.

The time-dependent temperature rise of a single component follows an exponential function with a time constant. When multiple layers are involved, there can be a number of different time constants working altogether to create a complex time response. Anyhow, assuming one time constant dominating in the transient thermal response, the time constant τ [1/s] is primarily determined by the heat capacitance C [J/K] and the thermal resistance ψ [K/W] of the system like in the resistor–capacitor (RC) circuit model used in an electric system, such that

$$\tau = C\psi \tag{4.25}$$

$$T(t)/T_{t\to\infty} = 1 - \exp\left(\frac{-t}{\tau}\right) \tag{4.26}$$

The time constant indicates the time taken to reach $(1-e^{-1}) = 63.2\%$ of the steady-state temperature for a step input. This relationship comes from the 1D heat diffusion equation, which is found in a usual heat transfer textbook.

A few earlier studies reported, such as Miner *et al.* [27] and Yang *et al.* [28], on transient thermoelectric cooling in the late 1990s or early 2000s. According to the more recent report by Ezzahri *et al.* [29], the transient temperature change in a microscale thermoelectric refrigerator was experimentally observed (Fig. 4.21) using a non-contact transient thermoreflectance imaging (TRI) technique [30,31]. The detailed information about TRI is provided in Chapter 7. The time resolution of the thermal imaging was a few microseconds. The microrefrigerator consisting of $Si/Si_{0.7}Ge_{0.3}$ superlattice layers (3 μm thick in total) was deposited on a silicon substrate with a top metallization layer of Ti/Al/Au (~ 2 μm thick). The thermal time response to a step current injection obtained from TRI is found in Ref. [32]. The thermal quadrupole method [33] was used for the modeling and analysis of the transient thermal response.

Depending on the thermal structure, a 1D model can be conveniently used if the device is just made of layers of materials. The first-order model can be constructed with the thermal mass model consisting of Joule heating and Peltier cooling. In this case, the thermal time response can be derived as

$$T(t) = \left(1 - \exp\left(\frac{-t}{\tau_J}\right)\right)(\psi_{TE} + \psi_P)Q_J$$

$$- \left(1 - \exp\left(\frac{-t}{\tau_P}\right)\right)(\psi_{TE})Q_P \tag{4.27}$$

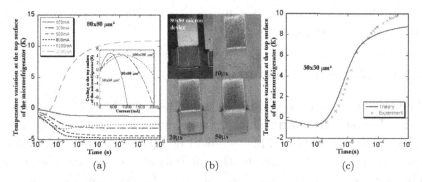

Fig. 4.21: Time response of temperature change by a step current to the thermoelectric device on silicon substrate. (a) shows the experimental time response, (b) shows the thermal images, and (c) shows the comparison of the developed model and measured data.

Source: Reproduced with permission from *J. Appl. Phys.*, **106**(11), 114503 (2009).

where Q_P in the second term is the cooling power by the Peltier effect (STI), hence the sign of the term is negative, τ_J and τ_P are the time constants due to the Joule heating and the Peltier cooling, respectively. The Q_J in the first term is the Joule heating, which acts on the thermal mass. The thermal mass is defined by the thermal resistance to the temperature reservoir and the heat capacitance of thermoelectric material as a volume effect. Joule heating happens relatively slower than Peltier effect, so $\tau_J > \tau_P$.

At an early time after the bias is applied, the temperature change in the substrate is sufficiently slow, while the cooling of the target rapidly takes place. In less than a few tenths of a second, the cooling happened first and then the temperature increased back and reach the steady state due to the slower joule heating as the drive current is kept on constant.

4.9 Vapor Compression Refrigeration and Thermoelectrics

For a refrigeration device, the COP is an important metric to evaluate/predict the operation cost. In this section, three important thermodynamic cycles are briefly revisited and compared to see the characteristic differences in COP.

4.9.1   *Gas refrigeration cycle*

The gas refrigeration cycle uses only the gas phase of a working fluid. The simplest form of the gas refrigeration cycle is a reverse cycle of Brayton cycle. The technical detail of the reverse Brayton cycle is found everywhere in thermodynamic books. As shown in Fig. 4.22, each component has the same function as those of the Brayton cycle, but heat flow is in the opposite direction. The compressor in the refrigeration cycle is required to provide a sufficient mechanical work to bring the heat as total enthalpy of the working fluid to heat exchanger 1. This heat exchanger is sometimes called a condenser following the similarity with a vapor compression refrigeration cycle discussed in the following section but does not condense anything in this case. The mechanical work to drive the compressor is supplied by the electromagnetic motor with an electrical power input. The compressor has an isentropic efficiency of approximately 65–75%, while that of the electrical-to-mechanical conversion is much higher at 90–95%. After the compressor, the heat is rejected through a heat exchanger 1, where some fraction of the enthalpy is going through. As the working fluid goes though the expander turbine, some enthalpy is recovered using the remaining fraction of the pressure. The expander turbine also has an isentropic

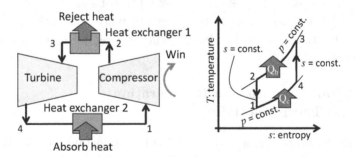

Fig. 4.22: Cycle diagram and T–s diagram of an ideal gas refrigeration cycle (the reverse Brayton cycle).

efficiency of approximately 70%. Then the mechanical power brings it back into the compression through the shaft. Typically, the gas cycle is a closed system. However, in special cases, such as in aircraft, ambient cold air can be used as a cold reservoir, hence it becomes an open cycle called the air cycle.

4.9.2   *Vapor−liquid two-phase refrigeration cycle*

In the vapor−liquid two-phase cycle, a liquid or liquid−vapor mixture phase of refrigerant absorbs the heat at the cooling side heat exchanger (evaporator) and transforms it into the latent heat to increase the vapor quality x: $x = 0\%$ while in a saturated liquid, $x = 100\%$ for pure vapor. Hence, the cycle can typically handle a larger amount of heat compared to the gas cycle. This cycle is also called the vapor compression cycle (VC) or vapor compression refrigeration cycle. In addition to the latent heat, some super heat by sensible heat is anticipated during the evaporation process. After compression, the refrigerant is cooled by external air through a heat exchanger (condenser), as shown in Fig. 4.23. In the condenser, vapor rejects the heat through the condensation process, giving away the latent heat in addition to the sensible heat. At the exit of the condenser, the working fluid is expanded through a thermally insulated expansion valve; this is an isenthalpic process in which the enthalpy of the system does not change, while the temperature of the working

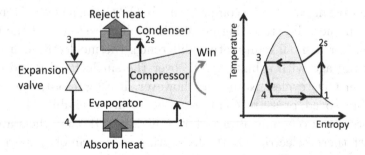

Fig. 4.23:   Cycle diagram and T–s diagram of typical vapor compression cycle with using expansion valve. Hatched area represents the liquid−vapor mixture state, while the left side of the border is liquid and the right side of the border is vapor.

fluid is reduced as the pressure decreases by expansion. This low-temperature liquid (or liquid—vapor mixture) can absorb the heat from the cooling target.

4.9.3 CO_2 trans-critical cycle and COP comparison

One of the unique characteristics of the carbon dioxide (CO_2) trans-critical cycle is that the critical point temperature is slightly above the room temperature. Hence, the cycle can use a trans-critical phase of CO_2 in the condenser for discharge compressed fluid in the cycle, hence a high heat transfer rate is expected. Eventually, the condenser heat exchanger can be made compact. However, the pressure needed to form the cycle is much higher than other refrigerants. Another challenge for this process is the condensation temperature. If the air temperature reaches a critical point ($31.1°C$), the performance becomes quite low. Thus, it is suitable to utilize this process in a colder climate zone. Usually, such air source cycles are considered to demonstrate a seasonal average performance when compared to the others. In this case, the intercooler is considered to recover heat before compression by reducing the condensation temperature.

Figure 4.24 shows the comparison of COP as a function of super-cooling temperature at the cooling target relative to the external temperature for three different isentropic efficiencies of the compressor in a CO_2 trans-critical cycle. The efficiency is determined by the discharge pressure at a given flow rate and the controlled temperature of the working fluid. The comparison of COP shows a clear difference between the solid-state thermoelectric cooling system and the CO_2 cycles. For a small temperature difference, thermoelectric can show an advantage with a higher COP over the single/fixed speed compression CO_2 cycles. Note that, however, if a state-of-the-art fully variable speed/pressure compressor is ideally available, the vapor compression cycle may always show a higher COP than the state-of-the-art thermoelectrics. Both the cutting-edge technologies are still under development, and it may be safer to mention that the thermo-electric cooling/refrigeration will perform well in a smaller temperature difference scale, where the compressors does not scale well to such a small cooling due to the impact of the volume-to-surface ratio.

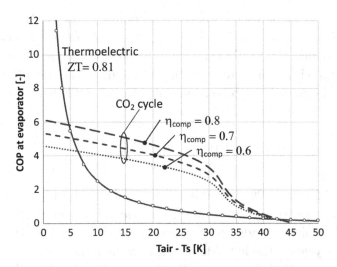

Fig. 4.24: COP as a function of heat pump temperature difference, e.g. pumping heat from 5°C to variation of air temperature. The trans-critical CO_2 refrigeration cycle with a fixed speed compression was investigated for three variations of isentropic efficiencies. The thermoelectric cooling is with constant Z value of 2.88×10^{-3} (1/K). For a small temperature difference, $ZT \sim 0.81$.

As another minor advantage for thermoelectric cooling, the COP curve shown for TE cooling in Fig. 4.24 is essentially independent of the absolute temperature, as the material figure of merit (ZT) can remain about the same even at deep cryogenic temperatures. The vapor compression cycles, however, will be greatly influenced by the absolute temperature range.

4.10 Thermoelectric Hybrid Cycle

As previously discussed, the CO_2 trans-critical cycle is desirable in terms of its environmental impact, including low CO_2 emission and low consumption of electrical power, which eventually saves the consumption of fossil fuel as well. The challenge for the cycle, however, is the relatively hot air temperature. One possibility is to use a thermoelectric device as a sub-cooler in the cycle or use the thermoelectric to pump out the heat to lower the temperature enough before returning the vapor back into the valve, so that nearly 100% of the latent heat can be used to boost the cooling capacity. In this case, interestingly, heat pumping can bring down the hot side temperature as well. Hence, the COP of the TE sub-cooler component can remain quite high with a small temperature gradient across the TE module. Due to the additional power consumption of a TE heat pump, the system COP must be redefined as

$$\text{COP} = \frac{q}{W_{\text{comp}} + W_{\text{TE}}} \tag{4.28}$$

Figures 4.25 and 4.26 show the diagrams of the hybrid cycle and the *P–h* characteristics as the thermoelectric sub-cooler is introduced. The sub-cooler extends the enthalpy removal larger than the simple baseline cycle on the left side of the *P–h* diagram.

By connecting the thermoelectric analytic model to thermodynamic calculation of the trans-critical CO_2 cycle [34], the system COP is calculated as a function of the TE sub-cooling temperature, which can be controlled by the current applied to the TE module (Fig. 4.27).

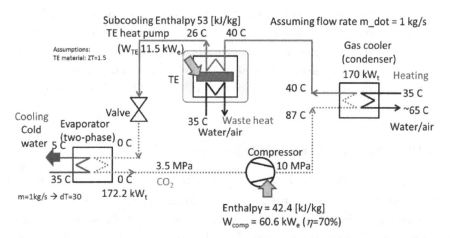

Fig. 4.25:   Cycle diagram of a thermoelectric–CO_2 hybrid cycle. In this example, thermoelectric works as a sub-cooler in the CO_2 cycle. The hot air condition 35°C is a disadvantage for the CO_2 cycle.

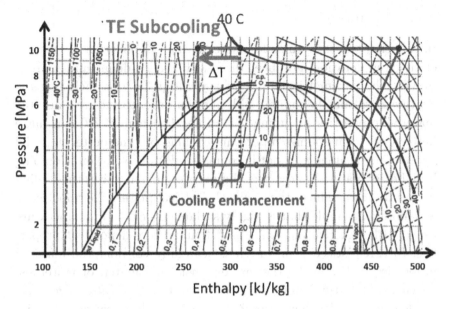

Fig. 4.26: *P–h* diagram of the cycle shown in Fig. 4.25, showing how the thermoelectric sub-cooler enhances the system performance.

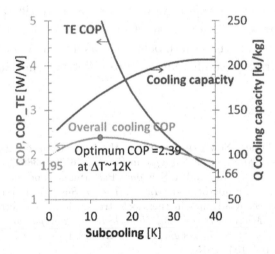

Fig. 4.27: COP of the hybrid cycle as changing sub-cooling temperature by thermoelectric. There exists an optimum operation point, where system COP increases by 2.39 compared to non-hybrid cycle at the hard condition for CO_2 cycle.

4.11　Summary

In this chapter, the principle, modeling, and application of thermo-electric refrigeration have been discussed. The modeling and analysis provide the parametric impact of key design factors, including the leg thickness and the number of legs per unit footprint area, and the material properties including ZT value. The higher ZT can make better COP near room-temperature cooling, but the better material figure of merit alone may not become the ultimate solution. Combing both better material and better thermal design maximizes the cooling performance or reduces the required electrical power. Thermoelectric is a unique principle that is highly scalable over a wide temperature range, although real materials have temperature-dependent proper-ties. In a realistic design, one must carefully consider the material properties and also the thermal expansion mismatch of materials, which could cause significant interfacial stresses. Multi-stage ther-moelectric modules can potentially provide a deep cryogenic refrig-eration in a very small scale such as hotspots, but it may not be suitable to cool the entire device structure. Comparison in COP has been made with vapor compression-based refrigeration cycles, which highlights the unique characteristics of thermoelectric refriger-ation. It is obvious that thermoelectric has a great benefit for appli-cations requiring a small temperature difference and small cooling features.

References

[1] J. C. A. Peltier, New experiments on the heat effects of electric currents, *Annales de Chimie et de Physique* in French, **56**, 371–386 (1834).

[2] Y. R. Koh, K. Yazawa and A. Shakouri, Performance and mass optimization of thermoelectric microcoolers, *Int. J. Therm. Sci.*, **97**, 143–151 (2015).

[3] P. Wang, Recent advance in thermoelectric devices for electronics cooling, *Cooling of Microelectronic and Nanoelectronic Equipment: Advances and Emerging Research*, M. Iyengar, K. J. L. Geisler, B. Sammakia (Eds.), World Scientific, 167–194 (2015).

[4] B. Pangrle, Power-Efficient Design Challenges, In: *Chips 2020 The Frontiers Collection*, B. Hoefflinger (Ed.), Springer, Berlin, Heidelberg (2011).

[5] A. Majumdar, Thermoelectric devices: Helping chips to keep their cool, *Nat. Nanotechnol*, **4**(4), 214 (2009).

[6] K. Yazawa and A. Shakouri, Optimum design and operation of thermoelec-tric heat pump with two temperatures, *ASME Interpack 2015* (2015).

[7] H. J. Goldsmid, Conversion efficiency and figure-of-merit [Chap. 3], In *CRC Handbook,* D.M. Rowe (Ed.), 19–26 (1995).

[8] W. M. Haynes (ed.) *CRC Handbook of Chemistry and Physics* (92nd ed.) Boca Raton, FL: CRC Press, 4.52 (2011).

[9] Y-M. Lin and M.S. Dresselhaus, Thermoelectric properties of superlattice nanowires, *Phys. Rev. B,* **68**(7), 075304 (2003).

[10] A. R. Abramson, W. C. Kim, S. T. Huxtable, H. Yan, Y. Wu, A. Majumdar, C.-L. Tien and P. Yang, Fabrication and characterization of a nanowire/polymer-based nanocomposite for a prototype thermoelectric device, *J. Microelectromech. Sys.*, **13**(3), 505–513 (2004).

[11] H. Fang, T. Feng, H. Yang, X. Ruan and Y. Wu, Synthesis and thermoelectric properties of compositional-modulated lead telluride–bismuth telluride nanowire heterostructures, *Nano Lett.*, **13**(5), 2058–2063 (2013).

[12] M. S. Dresselhaus, G. Chen, M. Y. Tang, R. G. Yang, H. Lee, D. Z. Wang, Z. F. Ren, J.- P. Fleurial and P. Gogna, New directions for low-dimensional thermoelectric materials, *Adv. Mater.*, **19**(8), 1043–1053 (2007).

[13] D. Bourgault, C. Giroud-Garampon, N.Caillault and L.Carbone, Thermoelectrical devices based on bismuth-telluride thin films deposited by direct current magnetron sputtering process, *Sens. Actuat. A: Phys.*, 273, 84–89 (2018).

[14] S.-D. Kwon, B.-K. Ju, S.-J. Yoon and J.-S. Kim, Fabrication of bismuth telluride-based alloy thin film thermoelectric devices grown by metal organic chemical vapor deposition, *J. Electron. Mater.*, **38**, 920–924 (2009).

[15] R. Venkatasubramanian, E. Silvola, T. Colpitts and B. O'quinn, Thin-film thermoelectric devices with high room-temperature figures of merit, in *Materials for Sustainable Energy: A Collection of Peer-Reviewed Research and Review Articles*, Nature Publishing, 120–125 (2011).

[16] H. Bottner, G. Chen and R. Venkatasubramanian, Aspects of thin film superlattice thermoelectric materials, devices, and applications, *MRS Bull.*, **31**(3) 211–217 (2006).

[17] G. Min, D. M. Rowe and F. Volklein, Integrated thin film thermoelectric cooler, *Electron. Lett.*, **34**(2), 222–223 (1998).

[18] V. Sahu, A. G. Fedorov, Y. Joshi, K. Yazawa, A. Ziabari and A. Shakouri, Energy efficient liquid-thermoelectric hybrid cooling for hot-spot removal, *The 28th Annual IEEE Semiconductor Thermal Measurement and Management Symposium (SEMI-THERM), IEEE*, 130–134 (2012).

[19] A. Shakouri, Nanoscale thermal transport and microrefrigerators on a chip, *Proceedings of the IEEE*, **94**(8), 1613–1638 (2006).

[20] R. Yang, G. Chen, G. J. Snyder and J.-P. Fleurial, Multistage thermoelectric microcoolers, *J. Appl. Phys.*, **95**(12), 8226–8232 (2004).

[21] https://www.amstechnologies-webshop.com/multistage-thermoelectric-peltier-modules-sw10283.

[22] G. Karimi, J. R. Culham and V. Kazerouni, Performance analysis of multistage thermoelectric coolers, *Int. J. Refrig.*, **34**(8), 2129–2135 (2011).

[23] R. Venkatasubramanian, Cascade cryogenic thermoelectric cooler for cryogenic and room temperature applications, U.S. Patent 6722140, issued April 20 (2004).

[24] F. D. Rosi, B. Abeles and R.V. Jensen, Materials for thermoelectric refrigeration, *J. Phys. Chem. Solids.*, **10**(2–3), 191–200 (1959).

[25] H. Goldsmid, *Thermoelectric Refrigeration 1964 Edition*, Springer, pp. 74–81 (1959).

[26] A. Ziabari, Ephraim Suhir and A. Shakouri, Minimizing thermally induced interfacial shearing stress in a thermoelectric module with low fractional area coverage, *Microelectron. J.*, **45**, 547–553 (2014).

[27] A. Miner, A. Majumdar, U. Ghoshal, Thermo-electro-mechanical Refrigeration Based on Transient Thermoelectric Effects, *Proc. 18th International Conference on Thermoelectrics*, 27–30 (1999).

[28] R. Yang, G. Chen, A. R. Kumar, G. J. Snyder, J-P. Fleurial, Transient cooling of thermoelectric coolers and its applications for microdevices, *Energy Convers. Manag*, **46**, 1407–1421 (2005).

[29] Y. Ezzahri, J. Christofferson, G. Zeng and A. Shakouri, Short time transient thermal behavior of solid-state microrefrigerators, *J. Appl. Phys.*, **106**(11), 114503 (2009).

[30] G. Tessier, S. Hole and D. Fournier, Quantitative thermal imaging by synchronous thermoreflectance with optimized illumination wavelengths, *Appl. Phys. Lett.*, **78**(16), 2267–2269 (2001).

[31] J. Christofferson, Y. Ezzahri, K. Maize and A. Shakouri, Transient thermal imaging of pulsed-operation superlattice micro-refrigerators, *Proc. 25th IEEE Semi-Therm Symposium*, 45–49 (2001).

[32] Y. Ezzahri, J. Christofferson, K. Maize and A. Shakouri, Short time transient behavior of SiGe-based microrefrigerators, *MRS Online Proceedings Library Archive*, 1166 (2009).

[33] D. Maillet, *Thermal Quadrupoles: Solving the Heat Equation Through Integral Transforms*, John Wiley & Sons (2000).

[34] K. Yazawa, S. Dharkar, O. Kurtulus, E. A. Groll, Optimum design for thermoelectric in a sub-cooled trans-critical CO_2 heat pump for data center cooling, *Proceedings of the Annual IEEE Semiconductor Thermal Measurement and Management Symposium*, 19–24 (2015).

Chapter 5

Power Generation

This chapter describes detailed modeling and analysis of thermo-electric (TE) power generation for various scales of application from a single thermoelectric element and modules to systems. Starting with the working principles, we discuss the design optimization of thermoelectric modules in contact with hot and cold thermal reservoirs. Due to the inter-play between the electrical and thermal energy transport, impedance matching for both electrical and thermal resistances are required to simultaneously achieve the maximum power output. Electrothermal optimization of thermoelectric elements with external contacts allows for a cost-effective and compact design for obtaining the desired power output. With leg thickness and fill factor as two important key variables, strategies for cost-effective design for TE power generators are discussed. Then, energy payback analysis with fluid pump power taken into account for the efficient convective heat exchange at both the hot and cold sides is provided. Theoretical maximum efficiency and power output are further discussed with asymmetric thermal contacts in the context of Carnot and Curzon−Ahlborn limits. Finally, impacts of temperature-dependent material properties, and the thermal stresses caused by large thermal gradients in TE generator modules during operation are briefly discussed.

5.1 Working Principle

A thermoelectric power generator is composed of thermoelectric elements, electrodes, thermally conductive and electrically insulating substrates, and lead wires. In a rigid module, two flat substrates

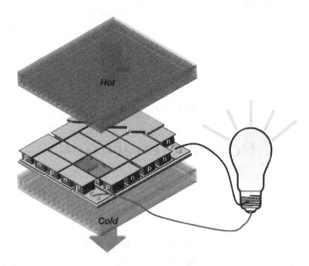

Fig. 5.1:   A schematic diagram showing the internal structure of a thermoelectric power generator with two external heat exchangers. The hot and cold side heat exchangers draw the heat from a hot medium flow and drain the waste heat to the cold side. Due to the temperature gradient across the thermoelectric device, an electrical potential is created between the two electrical terminals, and an electric current flows in the closed circuit with an external load (light bulb), where a useful work is extracted.

sandwich the internal structure and make external thermal contacts on both sides. Inside the module, an array of the n-type and p-type thermoelectric material pairs, called legs, are alternatingly connected thermally in parallel and electrically in series with electrodes. The layout usually follows a checkerboard pattern. Figure 5.1 schematically shows the internal structure of a thermoelectric power generator along with external heat exchangers. Both hot and cold side heat exchangers are used to effectively transfer heat, from heat reservoir to hot side of the module and cold side to cold reservoir, respectively. By converting heat into an electrical current within the module, power is extracted at an external electrical load in a closed circuit. In this figure, the power output is drawn to light a light bulb as an example.

Let's take a closer look at a single thermoelectric element (leg). A leg is typically made of a semiconductor material. When a temperature difference is applied between the top and bottom of the leg, a potential $S(T_h - T_c)$ [V] is created across the leg. The coefficient S [V/K] is called the Seebeck coefficient, named after Thomas Johann Seebeck, and is also known as the thermopower. The coefficient is an intrinsic material property that does not depend on the size of the leg. The Seebeck effect is caused by the diffusion of charged particles (carriers), which are negatively charged electrons in the case of n-type or positively charged holes in the case of p-type. Carriers travel across the confined volume of a solid matter. In this case, the solid is a thermoelectric leg. In equilibrium, the average kinetic energy of carriers proportionally increases with temperature by receiving thermal energy from the lattice. Due to the imbalance in their kinetic energy, both types of carriers tend to diffuse from higher to lower temperature, resulting in a voltage development across the leg. By connecting the p-type and n-type element pair with an electrode and to an external load resistance R_L, the developed voltage creates a current I flows in the same direction through the legs. The legs are electrically connected in series and thus an electric power is generated and delivered to the load in the circuit. The generated power is equal to the power consumption at the load, i.e. $I^2 \times R_L$[W]. Note that the electrical current flows along the n-type and p-type elements in the opposite directions. Figure 5.2 illustrates this phenomenon.

The temperature gradient is maintained by the thermal resistance of the leg, which is determined by the thermal conductivity and

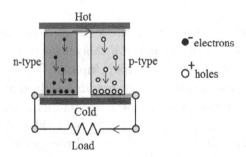

Fig. 5.2: A pair of n-type and p-type elements in contact with temperature reservoirs. The circuitry is closed by connecting the n-type and p-type elements to a load resistor. Due to the voltage created in the elements by the Seebeck effect, an electric current flows in the circuit, and thus, the power is delivered to the load.

dimensions of the leg. Similarly, the electrical resistance is determined by the electrical conductivity and the dimensions. A thermoelectric generator can be modeled as a battery in an electrical circuit with these resistances as the internal resistances of the battery. The difference between a TE generator and an ordinary battery is that thermoelectric legs include both electrical and thermal transport simultaneously while a battery includes only electrical transport. The two material properties, the electrical and thermal conductivities, are two crucial factors that determine the performance of a thermoelectric material along with the Seebeck coefficient.

When the circuit is open, there is no power output because no current goes through the load resistance. Nevertheless, there is an electrical potential created across the device. This is called the open circuit voltage (V_{OC}). This occurs due to a combination of the Seebeck effect at the terminals and the Thomson effect inside the element. When the circuit is closed, Joule heat is generated in the legs by the current flow through the internal resistance. Also, when current flows, a Peltier effect occurs at the junction (terminal). Quantitative discussion and modeling with all these different heats involved will be discussed in a later section.

Thermoelectric elements are generally made of semiconductors, but as long as the material conducts heat and electric current with a certain level of Seebeck coefficient, it can be a potential option, which includes metal conductors, conducting polymers, or semi-insulator ceramics.

Temperature gradient across the leg is an essential condition for power generation. In the model of a thermoelectric power generator system, instead of fixing the temperatures of the leg terminals, the temperatures of the two thermal reservoirs are fixed and connected: a high temperature (heat source) and a low temperature (heat sink). In this case, the terminal temperatures on both sides of the leg change with the heat flow Q [W] and the thermal resistance ψ [K/W] of the heat transfer component involved.

In this discussion, all phenomena are assumed to be in the thermal equilibrium, i.e. the charge carriers and the local lattice are at an equal temperature. Even for the transient model dealing with time-dependent temperatures, the heat flow must be in quasi-equilibrium over time. It is also assumed that the thermal energy associated with each thermal reservoir has an infinitely large capacity.

5.2   Design Optimization for Power Generation

Much of the research efforts on thermoelectric materials has focused on improving the dimensionless figure of merit (ZT) of the material [2–4]. The material figure of merit (Z) excluding the temperature component from ZT is composed of the Seebeck coefficient S [V/K], the electrical conductivity σ [S/m], and the thermal conductivity β [W/(m K)], such that

$$ZT = \frac{\sigma S^2}{\beta} T \qquad (5.1)$$

The numerator σS^2 is also called the thermoelectric power factor. Majority of the studies on improvement of the figure of merit has been directed toward reducing the thermal conductivity, for instance, by superlattices or embedded nanoparticles, while keeping the power factor unaffected. There have been several other studies on improving the power factor through the band engineering [5,6] and electron energy filtering by heterostructures [7, 8]. Approaches for improving the material figure of merit is discussed in the following sections. Realistically, the material properties constituting ZT are all temperature-dependent, such that

$$Z(\theta)T = \frac{\sigma(\theta) S(\theta)^2}{\beta(\theta)} T, \; \left(T - \frac{\Delta T}{2} \leq \theta \leq T + \frac{\Delta T}{2}\right) \qquad (5.2)$$

where T is the average temperature and ΔT is the temperature difference applied. Figure 5.3 shows examples of silicon germanium (SiGe) thermoelectric materials on their temperature-dependent ZT values and theoretical calculations by the Boltzmann Transport Equation Solver (BTEsolver) simulation tool. More details about the BTEsolver tool are found in Chapter 9.

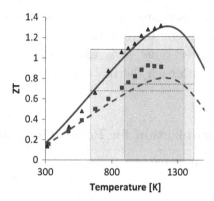

Fig. 5.3:   Temperature-dependent ZT values of n-type SiGe by Wang *et al.* [14] (square dot) and p-type SiGe by Joshi *et al.* [15] (triangle dot) with theoretical fitting curves by the open access Boltzmann solver developed by Je-Hyeong Bahk *et al.* [16]. The taller box shows the temperature band (905–1422 K) of high-temperature utilization and the mean ZT values are 0.75 and 1.21 for n-type and p-type, respectively. The lower box shows the full temperature band (645–1335 K) assuming a cold reservoir temperature 300 K, where the mean ZT values are 0.68 and 1.08 for n-type and p-type, respectively. Both use the heat source temperature 1680 K.

The ZT value is not the only factor affecting the power output. As aforementioned, the heat-to-electricity energy conversion process involves thermal contacts with hot and cold reservoirs. The system efficiency at maximum power output is inversely proportional to the sum of the heat dissipation at the hot and cold sides [9]. Optimal conditions are only found if the internal thermoelectric impedance matches the external impedance both electrically and thermally. This fact has been reported recently in the literature on thermoelectric systems [10]–[13]]. To understand how it works, a simplified analytical model is developed. The model includes external heat transfer that reflects the module design, materials, and conditions.

5.2.1   *Analytic model*

Figure 5.4 shows the thermal and electric circuit models for a single TE leg generator. It also shows how heat and electrical energy flows in and out. The model consists of a thermoelectric element

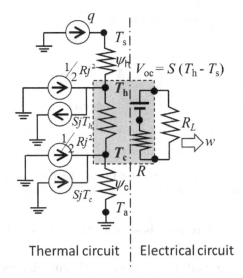

Thermal circuit ¦ Electrical circuit

Fig. 5.4: Electrothermal resistance network model used for a single-leg thermo-electric generator. A thermoelectric leg is in contact with hot and cold reservoirs as well as the external electrical load. In the thermal circuit, heat that comes in and out by the Peltier, Seebeck, and Joule effects are also shown.

(leg) of length d placed between hot and cold reservoirs. For simplicity, we assume that all material properties of p-type and n-type legs are constant and have the same value except for the opposite sign of Seebeck coefficients. Models and simulation with temperature-dependent material properties are discussed in Chapter 9.

Thermal resistance of the hot reservoir is denoted as ψ_h [K m^2/W] and thermal resistance of the cold reservoir is denoted as ψ_c [K m^2/W]. This model considers a unit cross-section area of the TE leg perpendicular to the heat flow. There are n (number) of legs [1/m^2] virtually built in the unit area. Heat flux q_h [W/m^2] is supplied by the hot reservoir at temperature T_s (fixed value). The cold reservoir is at temperature T_a (fixed value). Heat flux q_c flowing into the cold reservoir is smaller than q_h as the difference is the power output. Useful power output w [W/m^2] per unit footprint area, which is an allocated substrate area to the array of legs, is extracted by the external electrical load resistance R_L [Ω/m^2] electrically connected

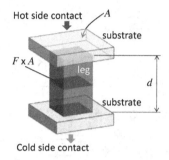

Fig. 5.5:   Geometrical representation of the model for a unit leg. F is the fill factor defined as the fractional area coverage by the TE leg.

to the leg. The footprint area is smaller than the total cross-sectional area of the legs in general.

For a particular material system, a thermoelectric power generator system that includes modules and hot side and cold side heat sinks (Fig. 5.1) can be designed for either maximum power output or maximum efficiency. Here, more focus is directed on the maximum power output because the power output asymptotically approaches zero when the efficiency approaches its maximum in a system with finite temperature hot junctions (heat sinks), as discussed later in this chapter.

Figure 5.5 reflects the physical representation of a unit leg, where an important parameter F called the fill factor has been introduced. The fill factor is defined as the fractional area coverage of a leg with respect to the footprint area of the substrate. F is less than unity. The inherent electrical and thermal characteristic, which is the conductance, of the leg proportionally changes with the fill factor.

Other important design parameters include thermoelectric leg thickness d [m], external thermal resistances ψ_h, ψ_c [K/W], and load resistance R_L [Ω]. The subscripts h and c represent the hot and cold temperatures of leg junctions, respectively. The subscript L indicates the load resistance. In practice, there is always non-zero thermal resistance between the leg and temperature reservoirs. The thickness and load resistance are also design variables to obtain the optimal output power. As mentioned earlier, changing the thermoelectric leg length affects the temperatures on the hot and cold sides and changes the flow of heat at the interface due to the Peltier and Seebeck effects. Therefore, the energy balance needs to be calculated recursively. Because of this complexity, the combined electrothermal network and mathematical processing must be carefully optimized. Solving the heat balance equations determines the parameters at the maximum power output.

Based on the energy conservation at all four temperature nodes shown in Fig. 5.4, the following equations are derived with T_s, T_h, T_c, and T_a. Along the direction of heat flow, following each equation, they are derived from left to right as heat flows in that order:

$$\frac{(T_s - T_h)}{\psi_h} = Q_h \tag{5.3}$$

$$Q_h = K\,(T_h - T_c) + SIT_h - I^2 R/2 \tag{5.4}$$

$$K(T_h - T_c) + SIT_c + I^2 R/2 = Q_c \tag{5.5}$$

$$Q_c = \frac{(T_c - T_a)}{\psi_c} \tag{5.6}$$

where

$$K = \frac{\beta F A}{d} \tag{5.7}$$

$$R = \frac{d}{\sigma F A} \tag{5.8}$$

K [W/K] is the thermal conductance of the leg, β is the thermal conductivity [W/(m K)], S is the Seebeck coefficient [V/K], I is the electrical current [A] induced by the thermopower, R [Ω] is the internal electrical resistance of the leg, and σ [1/(Ω m)] is the electrical conductivity. Q_h [W] and Q_c [W] represent the heat flow coming from the source reservoir and the heat flow going to the heat sink (cold reservoir), respectively. The first term on right-hand side of Eq. (5.4) represents the heat diffusion following the Fourier's law. The second term is the removal or addition of heat by the Peltier effect, and the third is Joule heating coming out of the TE leg. Note that Joule heat is generated everywhere inside the thermoelectric leg. In one-dimensional heat transport, the Joule heating is equally divided and go out to the two ends of the TE leg. The output power W [W] is found as the product of current I and the voltage drop across the load resistance using the resistance ratio m defined by $m = R_L/R$:

$$W = I(IR_L) = I^2 mR = \frac{I^2 md}{\sigma F A} \tag{5.9}$$

This electrical current I is determined by open circuit voltage $V_{oc} = (ST_h - ST_c)$ and the total electrical resistance of the circuit, such that

$$I = \frac{S(T_h - T_c)}{(R + R_L)} = \frac{\sigma FAS(T_h - T_c)}{(1 + m)d} \tag{5.10}$$

By substituting Eq. (5.10) into Eq. (5.9),

$$W = I^2 mR = \frac{m\sigma S^2 FA}{(1+m)^2 d}(T_h - T_c)^2 \tag{5.11}$$

To obey the system energy conservation,

$$W = Q_h - Q_c \tag{5.12}$$

Until here, the power output is a direct function of the temperature gradient $(T_h - T_c)$ across the leg. The interest of the relation, however, is in the power output induced by the overall temperature gradient $(T_s - T_a)$ instead. The ratio of these temperature differences, α, is introduced and determined from the energy balance equation as

$$\alpha = \frac{(T_s - T_a)}{(T_h - T_c)} = \frac{d + \beta FA(X + Y)}{d} = 1 + K(X + Y) \tag{5.13}$$

where X and Y are

$$X = \left(1 + \frac{Z}{2(1+m)^2}\left((2m+1)T_h + T_c\right)\right)\psi_h \tag{5.14a}$$

$$Y = \left(1 + \frac{Z}{2(1+m)^2}\left(T_h + (2m+1)T_c\right)\right)\psi_c \tag{5.14b}$$

where X and Y are the transformational external thermal resistances on the hot and cold sides, respectively. These are based on the true thermal resistances but modified by the addition and removal of the internal heat, which is induced by the electrical current flow. Using the above effective external thermal resistances X and Y, power output can be written as a function of T_s and T_a as

$$W = \frac{mZ}{(1+m)^2} \frac{\frac{\beta F A}{d}}{\left(1 + \frac{\beta F A}{d}(X+Y)\right)^2} (T_s - T_a)^2 \qquad (5.15)$$

The temperatures T_h and T_c in Eq. (5.14) can be found by recursively applying the following equations, which are transformed from the energy balance equations, Eqs. (5.4) and (5.5). In subsequent analysis, T_h and T_c are calculated iteratively:

$$g_1 = \frac{\beta}{d} X (T_h - T_c) - (T_s - T_h) = 0 \qquad (5.16)$$

$$g_2 = \frac{\beta}{d} Y (T_h - T_c) - (T_c - T_a) = 0 \qquad (5.17)$$

5.2.2 *Optimization for maximum power output*

To maximize the power output, the Lagrange multiplier method was used to find the optimal parameters for maximum power output using the independent parameters m, d, T_h, and T_c. Taking partial derivatives of Eq. (5.15) with respect to m, d, T_h, and T_c while introducing equality constraint functions g_1 and g_2 in Eqs. (5.16) and (5.17), the following eight equations are obtained:

$$\frac{\partial w}{\partial m} - \lambda_1 \frac{\partial g_1}{\partial m} = 0, \quad \frac{\partial w}{\partial d} - \lambda_1 \frac{\partial g_1}{\partial d} = 0, \quad \frac{\partial w}{\partial T_h} - \lambda_1 \frac{\partial g_1}{\partial T_h} = 0,$$

$$\frac{\partial w}{\partial T_c} - \lambda_1 \frac{\partial g_1}{\partial T_c} = 0, \quad \frac{\partial w}{\partial m} - \lambda_2 \frac{\partial g_2}{\partial m} = 0, \quad \frac{\partial w}{\partial d} - \lambda_2 \frac{\partial g_2}{\partial d} = 0,$$

$$\frac{\partial w}{\partial T_h} - \lambda_2 \frac{\partial g_2}{\partial T_h} = 0, \quad \frac{\partial w}{\partial T_c} - \lambda_2 \frac{\partial g_2}{\partial T_c} = 0 \qquad (5.18)$$

First, the Lagrange multiplier λ_1 and λ_2 are found from the two above Lagrange differentials of w with m as

$$\lambda_1 = \frac{(m-1)(T_h - T_c)}{\psi_h(mT_h + T_c)}, \quad \lambda_2 = \frac{(m-1)(T_h - T_c)}{\psi_c(T_h + mT_c)} \qquad (5.19)$$

Then, the optimum m, e.g. m_{opt}, is found by substituting Eq. (5.19) into λ_1 and λ_2 in the two above Lagrange differentials of w with respect to d. These two equations yield the same result as

$$m_{opt} = \sqrt{1 + Z\frac{(T_h + T_c)}{2}} \qquad (5.20)$$

This optimum m-value for the maximum power output is called 'thermoelectric factor' for impedance matching that is similar to the impedance matching factor of an electrical circuit for a battery. As is well known, the electrical impedance matching factor for a battery in DC mode must instead be unity. The discrepancies occur due to the interaction between the electrical and thermal transport. Here, the thermal circuit creates electrical irreversibility, and the electrical circuit inherently causes disruption of heat flow. Due to the thermoelectric energy conversion, the thermal resistance matching takes into account the reduction in the effective thermal conductance of the leg due to the electric current flow during power generation.

The optimal impedance factor shown in Eq. (5.20) can also be found in optimizing for the maximum efficiency in the case of the fixed temperatures at the contacts of the leg, T_h and T_c [17]. The following section discusses the efficiency of a system with finite and irreversible thermal contacts.

Note that in Eq. (5.20), this m factor is still dependent on T_h and T_c. From the rest of Lagrange derivative of w with respect to T_h and T_c, the leg length d of the maximum power output is obtained as

$$\frac{d_h}{\beta} = \frac{\psi_h(T_h + (2m-1)T_c)}{(T_h + T_c)}, \quad \frac{d_c}{\beta} = \frac{\psi_c((2m-1)T_h + T_c)}{(T_h + T_c)} \qquad (5.21)$$

where subscripts h and c denote the origin of the equations in the Lagrange differentials T_h and T_c, respectively. The solution of the optimum leg length is eventually obtained as the sum of the above equations as

$$\frac{d_{opt}}{\beta} = \frac{\psi_h \left(T_h + (2m-1)\, T_c\right) + \psi_c \left((2m-1)\, T_h + T_c\right)}{(T_h + T_c)} \tag{5.22}$$

The m in Eq. (5.22) must also obey Eq. (5.20), where m_{opt} replaces the m in Eq. (5.22).

Then, two temperatures T_h and T_c at the maximum power output are found from the given temperatures T_s and T_a. The ratio gives the factor that changes the terminal temperatures of the leg:

$$\alpha \equiv \frac{(T_s - T_a)}{(T_h - T_c)} = 1 + K(X + Y) \tag{5.23}$$

Then, substituting Eq. (5.22) into Eq. (5.23),

$$\alpha = \frac{(2m-1)(m+1)(\psi_h + \psi_c)(T_h + T_c) + 2(\psi_h(T_h + mT_c) + \psi_c(mT_h + T_c))}{(m+1)(\psi_h(T_h + (2m-1)T_c) + \psi_c((2m-1)T_h + T_c))} \tag{5.24}$$

This α represent the influence of asymmetric thermal contacts in power generation and the optimum leg design. In a special case with symmetric contacts, i.e. $\psi_h = \psi_c = \psi$, the optimum leg length becomes as $d_{opt}/\beta = 2m\psi$. This indicates that the optimal temperature difference across the leg for maximum power output must be a half of the total temperature difference, such that

$$\alpha_{symmetry} = \frac{T_s - T_a}{T_h - T_c} = 2 \tag{5.25}$$

Leg thickness d plays a significant role in order to obtain the maximum power output w. The optimum leg length d_{opt} is found by taking the derivative of w with respect to d in Eq. (5.15) to be zero, hence

$$d_{opt} = \frac{4}{\alpha^2} \kappa \beta \sum \psi \tag{5.26}$$

where κ is the ratio of the internal leg thermal resistance to the sum of the external thermal resistances $\sum \psi$, which is the sum of external thermal resistances. From the numerical tests with this analytic model, it is found that this factor κ must be the same as the

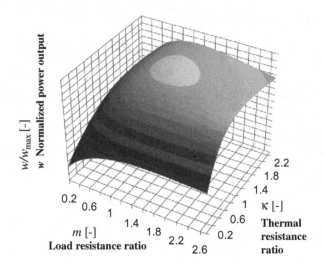

Fig. 5.6:   Normalized power output as function of electrical and thermal resistance ratios. Both electrical and thermal resistances match with the external resistances by the factor of $\sqrt{1 + ZT}$ at a unique point, where the power output achieves its maximum. In this figure, $ZT = 1$ (fixed at all conditions) with temperature at the contact are given (temperature reservoirs) as $T_s = 400$ K, $T_a = 300$ K.

electrical impedance ratio to obtain the maximum power output, $\kappa = m$. Figure 5.6 shows a sample case.

To consider the real dimensions of the legs, the actual cross-sectional area $F \times A$ is multiplied to find the real thickness, where A is the footprint area designated to the thermoelectric leg and F is the fill factor of the leg. Then, the formula is rewritten as

$$d_{opt} = \frac{4}{\alpha^2} m \beta F A \sum \psi \qquad (5.27)$$

For a special case of symmetric contacts with $\psi_h = \psi_c$, $\frac{4}{\alpha^2} = 1$, hence

$$d_{opt.symmetric} = m \beta F A \sum \psi \qquad (5.28)$$

This equation is valid only for the symmetric external resistance case, but the value of d_{opt} is observed very similar to this even for asymmetric cases since the sensitivity of thermal resistance ratio ψ_h/ψ_c to α is typically quite small. Therefore, Eq. (5.28) may be considered as a general optimal condition for thermoelectric power generators.

From Eq. (5.23) and (5.24), the relation of temperature ratio T_a/T_s and T_c/T_h is found as

$$\frac{T_a}{T_s} = \frac{\begin{array}{c}[\psi_c(2m(T_h + mT_c)) + (m+1)\{\psi_h(T_h + (2m-1)T_c)\\ + \psi_c((2m-1)T_h + T_c)\}]T_c - \psi_c(2m(T_h + mT_c))T_h\end{array}}{\begin{array}{c}[\psi_h(2m(mT_h + T_c)) + (m+1)\{\psi_h(T_h + (2m-1)T_c)\\ + \psi_c((2m-1)T_h + T_c)\}]T_h - \psi_h(2m(mT_h + T_c))T_c\end{array}}$$

(5.29)

Here, the temperatures T_c and T_h should be determined from Eqs. (5.16) and (5.17). However, the equations are still too complex to yield a closed-form solution, so these temperatures are retained in the formula. Finally, the maximum power output as a function of $(T_s - T_a)$ is found by modifying Eq. (5.11) with Eqs. (5.23) and (5.27) as

$$W_{max} = \frac{m\sigma S^2}{(1+m)^2} \frac{FA}{d_{opt}} (T_h - T_c)^2$$

$$= \frac{Z}{4(1+m)^2 \sum \psi} \frac{1}{(T_s - T_a)^2}$$

(5.30)

This equation is valid for any value of constant thermoelectric properties. In order to achieve this maximum power output per given condition and material, the optimum thickness of the leg must be designed to match Eq. (5.27). Looking into Eq. (5.30) closer, it can be seen that the first factor of the right-hand side consists of pure material properties, followed by a factor describing the thermal conduction of the contacts, and then a square of the entire temperature gradient of the system in the last. It is important to note that the maximum power output is inversely proportional to the total external thermal resistance and directly proportional to Z.

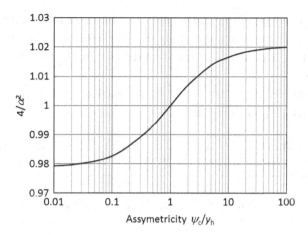

Fig. 5.7: Influence of the asymmetric external thermal resistance factor, ψ_c/ψ_h, on the thermal impedance matching, $4/\alpha^2$ for maximum power output. The plot is on the condition of $T_s = 400$ K, $T_a = 300$ K, $ZT = 1$.

In the special case of the symmetric contacts, Freunek *et al.* [18] derived the equation similar to Eq. (5.30):

$$W_{max} = \frac{Z}{4\sum\psi}(T_s - T_a)^2 \qquad (5.31)$$

Their model gives the maximum power output without the factor $(1 + m)^2$ in the denominator, and thus, their model is valid only if m is close to unity, implying $ZT << 1$.

The impact of asymmetricity in thermal contacts on α is quite small in practical cases. However, for the grasp of the trend, Fig. 5.7 shows a broad range of cases where the ratio of the two external thermal resistances varies from 0.01 to 100 while the material property of the leg remains constant, $ZT = 1$. The graph clearly shows that the assumption of factor $4/\alpha^2 \approx 1$ in Eq. (5.27) is quite robust for most of the practical designs regardless of the asymmetricity of thermal contacts.

5.2.3 *Efficiency analysis*

The energy conversion efficiency η at maximum power output is given by the power output W_{max} divided by the heat input Q_h. From

Eqs. (5.3), (5.11), (5.24), and (5.27), the system efficiency at W_{max} becomes

$$\eta = \frac{W_{\max}}{Q_h} = \frac{(m-1)(T_h - T_c)}{(mT_h + T_c)} \tag{5.32}$$

where m in the formula is derived from Eq. (5.20) replacing m_{opt} with m, namely, the thermoelectric factor. This equation is exactly the same as the well-known formula for the maximum efficiency of a single-leg thermoelectric generator when the terminal temperatures of the leg, T_h and T_c, are fixed and given. In the irreversible systems, T_h and T_c are variables controlled by the external thermal resistances and the leg dimensions.

5.3 Cost-Effective System Design

Due to the nature of power generators, the cost of a thermoelectric generator must be measured in cost per electric power output ($/W). An important strategy of cost-effective design for a thermoelectric power generator is to address the minimum material mass for maximum power output in the design. The idea behind the 'minimum material mass' comes from the basic concept of cost reduction, where it is assumed that the raw material market price per unit mass drives the majority of the device cost. As production scales up, the material cost begins to dominate the overall cost within the cost structure of manufacturing, including the cost of additional equipment and tools, administration cost, and so on. Except for improving the process and assembly, minimizing the material mass is a principal scientific issue that is directly translated into the cost-effective design. Because thermoelectric materials are relatively expensive, especially for those used in generator modules, the design needs to focus on reducing the material mass while maintaining the power output close to the maximum. For material selection, always 'the better the ZT is, the bigger the power output'. Improving the material (the ZT value) will certainly change the cost per power. Interestingly, however, this scenario does not always work to reduce the cost per power. The discussion begins with fixed material properties ($ZT = 1$ or equivalent) and considers the individual effects of the three component properties of a thermoelectric material, S, σ, and β, in the following sections.

5.3.1   *Key parameters*

As shown in the previous section, the optimum thickness of the leg for the maximum power output is given by

$$d_{opt} = \frac{4}{\alpha^2} m \beta F A \sum \psi \qquad (5.33)$$

The greater the thickness is, the higher the efficiency. The important factors impacting the cost are the leg cross-section FA and the thickness d. As F decreases, thickness d must decrease linearly to maintain the maximum power output. This means that the mass of material used for the leg decreases in proportion to F^2. On the other hand, to reduce F, there are limiting factors to consider, which are, among others, the thermal and electrical contact resistances.

5.3.2   *Thermoelectric module design*

TE modules discussed in this section are always optimized for the maximum power output of the system. A TE module consists of multiple thermoelectric legs, a hot side substrate, and a cold side substrate. Both substrates are assumed to have the same thickness d_s. Power output per mass is found by dividing the power output per total mass of the module. The mass of the module includes both types of TE legs with mass density of ρ_{TE} and the substrates with density of ρ_s. In a typical module geometry, the volume of contact materials (including electrodes) is significantly smaller than the mass of TE legs, so they are excluded from the total mass. Total mass per unit area is given by

$$M = \rho_{\mathrm{TE}} F d + 2\rho_s d \qquad (5.34)$$

As the substrate mass term is typically smaller than the TE mass with a module with conventional fill factor (e.g. 40%), the total mass decreases almost linearly as fill factor decreases. Here, two factors involved in heat transfer need to be carefully considered. These are the internal thermal resistance of the legs and the external contact thermal resistance. To simplify the problem, we ignore the parasitic heat loss through the space gap between the legs with a cross-sectional of $(1 - F)A$. The contact thermal resistance comes from the conduction of heat through the entire thickness of the substrate, including spreading or contraction thermal resistance. The heat spreading thermal resistance to both contacts of the leg and substrate increases as the fill factor is made smaller. The spreading resistance is determined by a model with closed formula [19], which is the geometry-dependent solution of the three-dimensional heat diffusion problem. Detailed discussion on heat spreading is found in Section 3.1.1 of Chapter 3.

Figure 5.8 shows the calculated maximum power output per mass as a function of fill factor for a sample TE generator. In the figure, h_h is the hot side heat transfer coefficient and h_c is the cold side heat transfer coefficient. The power per unit mass could reach thousands

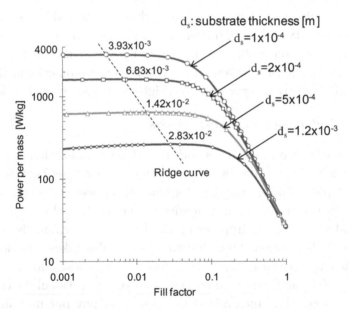

Fig. 5.8: Example of maximum power density per mass of module in variation of substrate thicknesses. Assumptions: $T_s = 600$ [K], $T_a = 300$ [K], $ZT = 1$, $\beta = 1.5$ [W/(m K)], $\beta_s = 140$ [W/(m K)], both h_h, and h_c are 500 [W/(m^2 K)]. Densities of TE material and aluminum nitride (AlN) substrate are 9.78×10^3 [kg/m^3] and 3.26×10^3 [kg/m^3], respectively.

of Watts/kg at low fill factors 0.3−3% compared to tens of Watts/kg for most of the off-the-shelf modules possessing large fill factors. The power output per mass remains nearly independent of the fill factor below 1%. At these low fill factors, the heat transport is significantly limited by the spreading resistances. For fill factors 1−10%, the substrate mass becomes dominant over the leg mass, so that the power per mass curves start to flatten. This result indicates the advantage of using a small fill factor (below 10%) leg and that one can essentially reduce the leg material by approximately 1/10,000−1/100 with very little performance degradation. Compared to conventional modules of their typical fill factors (e.g. 40%), the module costs may be reduced by 1/40 with a fill factor as small as 3%. For the module design with small fill factor legs, a material with high thermal conductivity and electrical insulation is desirable for the substrates, such as aluminum nitride (AlN), silicon carbide (SiC), or beryllium oxide (BeO).

These calculation results suggest that a material for substrate with low density yields a lightweight module as well. In this case, any thermal expansion coefficient mismatch between connected material must also be carefully taken into account as thermomechanical integration is important, especially for high-temperature applications.

5.3.3 *Net power and system cost*

Using an effective thermal network model, co-optimization of the TE module with heat sinks can be performed for various heat sources. The pump work power required for convection heat transfer to either side of the power generator is defined as the product of the mass flow rate and the pressure drop across the heat sink or channels by wall friction. This required fluid dynamic power also takes into account the efficiency of the pump. Then, this power is lost and should be subtracted from the generated power in order to calculate the net power output. The maximized net power output per unit area as a function of hot side heat flux is plotted in Fig. 5.9. While the reservoir temperatures T_s and T_a are fixed, the thermal resistances of the system vary. TE leg thickness is optimized for maximum power output for each of the three different fill factors, 0.01, 0.1, and 0.1.

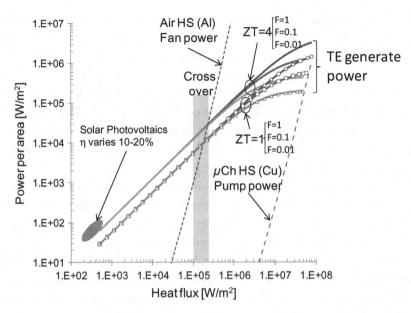

Fig. 5.9: Power output per unit area and pump power consumption for air heat sink and microchannel water heat sink at the optimum design: $T_s = 600$ K, $T_a = 300$ K, $ZT = 1$ with $\beta = 1.5$ W/(m K) and $ZT = 4$ with $\beta = 0.75$ W/(m K), Fan efficiency of 30%, pump efficiency of 60% and the efficiency of Solar Photovoltaic is assumed in the range of 10–20%.

Source: Reproduced with permission from *Environ. Sci. Technol.*, **45**(17), 7548–7553 (2011).

Straight rectangular channels are assumed for the flow passage of heat sinks either for air or water flow. The dimensionless heat transfer coefficient, Nusselt number, is determined by the channel cross-section geometry based on the uniform temperature of the wall. As channels are long enough relative to the size of channel, the flow in most of the sections would likely be fully developed. Hence, the Nusselt number is considered as a constant, while the assumption of laminar flow is not violated. The heat sink is assumed to be made of a thermally conductive material and the impact of fin efficiency is assumed to be negligibly small.

As heat flux increases, heat sink fin spacing needs to be decreased in order to extend the convective surface. Simultaneously, the flow rate must increase to be able to pump more heat. Tighter fins and higher coolant fluid velocity both require more pumping power, so that the pump power curve increases steeply as a function of heat flux. At some point, the electric power supplied to the pump overtakes the TE power output and net power output goes to zero. This happens, for instance, at around 10^5 W/m^2 heat flux for air cooling for this particular example. One can design water cooling channels for higher heat flux. Even such water cooling with microchannel technology will reach the limit at some point, where the pump power is required more than the generating power.

Such saturation point of trade-off depends on the fractional area coverage of the thermoelectric elements. The maximum input heat flux can be over 1 MW/m^2, which is rarely achieved. By comparison, the efficiency of the most commercialized photovoltaic is in the range of 10−20% and the solar power available without concentration is in the range of 700−1000 W/m^2. A concept of 'thermal concentration' in Chen *et al.* [21] for solar thermoelectric power generator gives a similar effect, while the mass requirement for the heat sink was not discussed.

The material cost per unit area for the same system is shown in Fig. 5.10. It can be seen that microchannel water cooling could cost less than air cooling at some of the lower heat fluxes (e.g. $\sim 10^4$ W/m^2 range), so the ultimate choice between air and water cooling should be decided based on the specific application. In our analysis, we only considered the cost of the key materials in a thermoelectric power generation system (TE material, substrate, and heat sink). The TE module manufacturing cost per unit area could be easily added to Fig. 5.10. In practice, one also has to consider the actual cost of the fan, water pump, the associated piping, etc. However, the analysis presented here provides a baseline to co-optimize the whole system and analyze the cost-efficiency trade-off for waste heat recovery applications. Manufacturing costs can be lowered as the production scales up by manufacturing learning curve. One will be ultimately limited by the cost of the raw materials in the system. Also, in comparison, Fig. 5.10 also plots the cost of poly-Si photovoltaic, which is around 400 $/m^2.

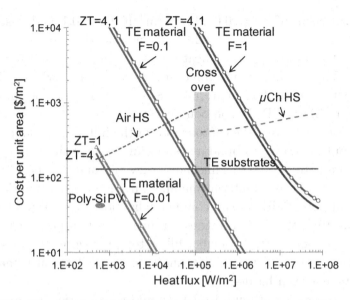

Fig. 5.10: The cost of materials for a thermoelectric system per unit area of the heat source: $T_s = 600$ K, $T_a = 300$ K, $ZT = 1$ with $\beta = 1.5$ W/(m K) and $ZT = 4$ with $\beta = 0.75$ W/(m K), fan efficiency of 30%, pump efficiency of 60%, TE at 500 \$/kg, aluminum nitride (AlN) substrate at the cost of 100 \$/kg, aluminum heat sinks at 8 \$/kg, copper microchannel at 20 \$/kg and poly-silicon (Poly-Si) photovoltaic (PV) material at 41 \$/m^2.

Source: Reproduced with permission from *Environ. Sci. Technol.*, **45**(17), 7548–7553 (2011).

Based on Figs. 5.9 and 5.10, if we use liquid cooling with inlet temperature T_a and 1×10^5 W/m^2 ($=10$ W/cm^2) as the input heat flux, then the net electrical power output will be 5.7×10^3 W/m^2 for $ZT_{\text{average}} = 1$ with an overall efficiency of 5.7% and 1.27×10^4 W/m^2 for $ZT_{\text{average}} = 4$ with an overall efficiency of 12.7%. The optimum leg length for this input heat flux is 32.7 μm for $F = 0.01$, 321 μm for $F = 0.1$ and 3.11 mm for $F = 1$. As expected, a smaller leg length can be used by reducing fill factor and matching thermal impedance with the heat source and the heat sink. The cross-section area of the leg is not linear to the leg length but has a strong relationship. Based on Fig. 5.10, the material cost for the heat sink will be 400 \$/m^2, for the two substrates, the cost will be 130 \$/m^2 and the leg will cost 1.34 \$/m² if $F = 0.01, 132$ \$/m² if $F = 0.1$ and costs will exceed 12,400 \$/m² if $F = 1$. This means that the overall 'material' cost for the TE module and heat sink will range from 0.095 \$/W to 2.27 \$/W depending on the fractional area coverage (when $ZT = 1$). The assumed cost of TE material at $\sim$500 \$/kg is based on the state-of-the-art BiTe or PbTe compounds.

Figure 5.11 shows the contour plot of the initial material cost per power output [\$/W] to build a system for different fill factors of (a) $F = 1$ and (b) $F = 0.1$. This shows how the material performance (the ZT value) and material manufacturing process (material unit cost) can potentially impact commercialization. With a fill factor as small as 10%, the TEG provides a range of cost to power as low as a few cents per Watt. At smaller fill factors, the slope becomes more sensitive to the ZT value. This indicates a need to push material development even harder.

By giving the optimum leg length and the electrical conductivity of the TE material, required minimum contact resistance is calculated. Desired range of parasitic Joule heating is set to less than 5% of overall performance, which is induced from the contact resistance. The minimum contact resistance needed for $F = 1$ is 3×10^{-5} [Ω cm], for $F = 0.1$ this is 3×10^{-6} [Ω cm], and for $F = 0.01$ this is 3×10^{-7} [Ω cm] while heat flux is 5×10^4 W/m² as an example. This highlights the importance of low contact resistivity in order to minimize the overall system cost. Another parasitic effect is the radiation from hot to cold plates for low fractional coverage areas, which we initially neglected. Estimates show that this can be negligible if we use low emissivity coating (e.g. 0.02 for gold coating). Air conduction inside the module is another source for heat leakage. If the thermal conductivity of the TE material is 1.5 W/(m K) at an average element temperature of 450 K, the thermal conductivity of the empty regions should be less than 0.001 W/(m K) for this parasitic heat path to be less than 5–6% of the total heat flow, when fill factor $F = 0.01$. When $F = 0.05$, this requires an air pressure of 4 Torr and an air conductivity of 0.006 W/(m K), which are calculated based on Potkay *et al.* [22] Considering the additional costs associated with vacuum packaging, one may decide to use larger fractional coverage, e.g. around $F = 0.05$.

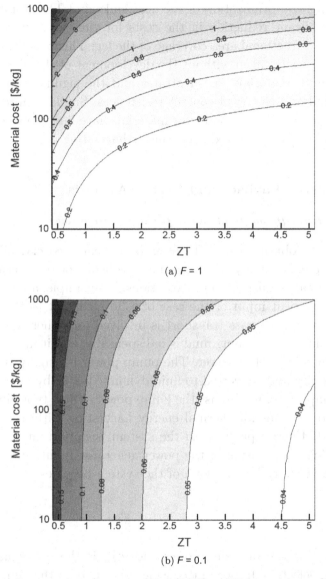

Fig. 5.11: Contour plot of cost per Watt [$/W] vs TE Material cost [$/kg] and ZT at maximum power: (a) $F = 1$ and (b) $F = 0.1$. Values on the curves indicate cost per Watt [$/W]. Darker regions show relatively higher prices for power and brighter regions show lower prices for power. The numbers on the curves show the value of power cost per Watt [$/W].

Source: Reproduced with permission from *Environm. Sci. Technol.*, **45**(17), 7548–7553 (2011).

Finally, co-optimization of the heat sinks and TE module allows for a significant reduction in the costs for the exotic materials by making low fractional area coverage of the leg. This benefit, however, cannot be easily taken without the development of the high thermal conductivity materials for substrates and the significant reduction of thermal and electrical contact resistances. Similar approaches to reduce the cost of semiconductor materials have been applied to tandem cells for concentrated solar energy harvesting.

5.4 Energy Payback and Exergy Analysis

5.4.1 *Energy payback with fluid pump power*

In order to obtain electrical power from heat recovery, additional power input is usually required since the heat flow is typically provided by mass transport. The exceptions, for example, are in the case of radiative heat input or the case of solid– contact to the thermal ground. As far as mass transport is involved for either the hot side or the cold side, pumping fluid is indispensable and it may require a non-negligible level of power. The pump power is usually supported by electricity and converted to fluid dynamic power by a mechanical fluid pump. Here, we discuss the pump power, which typically is necessary to consider for thermal energy harvesting with a fluid mass transport. Energy payback to the system is important to find the most efficient operation of the power generator taking into account the pump power. The net gain of the system becomes

$$w_{\text{sys}} = w_{\text{TE}} - w_{\text{pp}} \qquad (5.35)$$

where w_{pp} is the pumping power and w_{TE} is the power harvested by thermoelectric. Important consideration here is that w_{TE} is also a function of w_{pp}. The minimum pumping power w_{pp} required to obtain the desired heat transport is predicted by the formulations in Fig. 5.12.

From Ref. [23], the thermal resistance match between the temperature sensitive mass flow and the convection yields the near

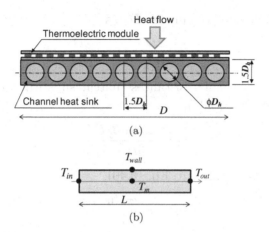

Fig. 5.12: Universal heat sink model. Part (a) shows the cross-section perpendicular to the flow passage and (b) shows the cross-section of the channel. The number of channels N varies depending on the device size $A = DL$.

maximum heat transport $U_B A_B$:

$$U_B A_B = 1 / \left(\frac{1}{2\rho C_p G} + \frac{1}{N U_{fin} A_{fin}} \right) \tag{5.36}$$

where U_B is the effective heat transfer coefficient from the device footprint area A_B. We look at the optimum thermal resistance match condition described as

$$2\rho C_p G = U_{fin} A_{fin} \tag{5.37}$$

Thus,

$$U_B = \frac{\rho C_p G}{DL} \tag{5.38}$$

The surface area A_{fin} of the convective heat transfer in a single channel is

$$A_{fin} = \pi D_h L \tag{5.39}$$

The heat transfer coefficient of the above area is found as

$$U_{fin} = \frac{\mathrm{Nu}\beta_f}{D_h} \tag{5.40}$$

where Nu is the Nusselt number and was found to be constant, Nu $= 4.634$ for circular channels [24]. The number of channels is defined by the channel diameter D_h as

$$N = \frac{2D}{3D_h} - 1 \approx \frac{2D}{3D_h} \tag{5.41}$$

The optimum thermal resistances match, substituting Eqs. (5.41) and (5.40) into Eq. (5.37),

$$G = \frac{2\pi \mathrm{Nu}\beta_f DL}{3\rho C_p D_h} \tag{5.42}$$

Substituting Eq. (5.42) into Eq. (5.40),

$$D_h = \frac{2\pi \mathrm{Nu}\beta_f}{3U_B} \tag{5.43}$$

The flow bulk velocity u is found as

$$u = \frac{4G}{N\pi D_h^2} \tag{5.44}$$

Assuming laminar flow and small contraction and expansion losses, the pressure drop across the channel is

$$\Delta P_{ch} = \frac{48\mu L}{D_h^2} u \tag{5.45}$$

Consolidating Eqs. (5.42)–(5.45), pumping power as a function of U_B is found as

$$w_{pp} = G\Delta P_{ch} = \frac{972\mu DL^3}{\pi^4 \left(Nu\beta_f\right)^3 \left(\rho C_p\right)^2} U_B^5 \tag{5.46}$$

Then, finally,

$$w_{\mathrm{pp}} \cong \frac{10\mu DL^3}{\left(\mathrm{Nu}\beta_f\right)^3 \left(\rho C_p\right)^2} U_B^5 \tag{5.47}$$

5.4.2   Case study for waste heat recovery

Here, an example case of waste heat recovery from a semiconductor chip is discussed. Heat source (junction) temperature $T_j = T_s$ and thermal ground temperatures T_a are constant, assuming both infinitely large thermal masses. Thermal resistance by the interface between the heat source and the hot side of TE is assumed constant

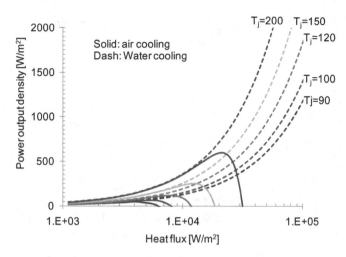

Fig. 5.13: Net power output per unit area vs input heat flux. The device area is 30 mm × 30 mm.

at $\psi_h = 0.01$ K/W. The ambient temperature is fixed to $T_a = 35°$C and the heat source temperature T_s is set to 90°C, 100°C, 120°C, 150°C, and 200°C. Temperature ratio T_a/T_s is in between 0.85 and 0.65. Three different scales are investigated as (a) 30 mm × 30 mm for a package, (b) 3 mm × 3 mm for a small chip, and (c) 300 μm × 300 μm for a hotspot.

Figures 5.13–5.15 show the net power output with respect to the heat flux of the source device. For the larger case (a), the harvesting levels are quite low while using air convection. Using water cooling, the power output curves are the same as in air convection but differ from the air convection at the higher heat flux condition. The pumping power needed for water cooling is more than four orders of magnitude smaller than that needed for air convection, based on Eq. (5.47), so that limitation is not observed until the heat flux reaches up to 10^7 [W/m^2] order. Higher temperatures yield better performance, as Eq. (5.47) clearly suggests. From the study, if the water cooling is available, energy payback is practical in up to 10^7 [W/m^2] order heat flux, and the higher T_j provides practical energy payback. Similarly, air cooling is not a good solution for the waste heat recovery for electronic devices.

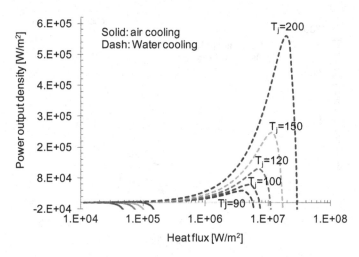

Fig. 5.14:   Net power output per unit area vs input heat flux. The device area is 3 mm × 3 mm.

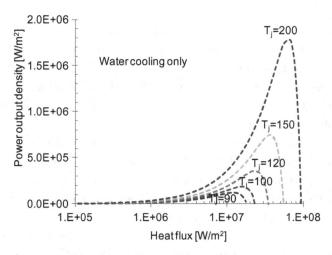

Fig. 5.15:   Net power output density vs input heat flux. The device area is 300 μm × 300 μm.

5.4.3   *Maximum power generation*

We will discuss entropy generation and the exergy of the waste heat recovery system in the following section. Prior to that, we focus on the maximum power generation first for later comparison. We took

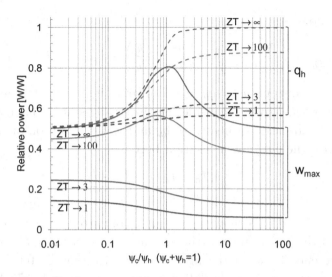

Fig. 5.16: Normalized power and heat by the maximum heat vs ψ_c/ψ_h with varying ZT. $T_a/T_s = 0.01$.

the liberty of varying the external thermal resistance to cover any case of thermal management. Figure 5.16 shows the power output and heat flow for different external thermal resistance ratios ψ_c/ψ_h, where the source temperature T_s and ambient temperature T_a are fixed.

In the case of $\psi_c + \psi_h = 1$ with $ZT = 1$, the maximum power output $w_{\max}$ is observed at the smallest ψ_c/ψ_h. Then the peak gradually shifts to a larger thermal resistance ratio. Finally, the peak approaches at $\psi_c = \psi_h$ as ZT goes to infinite. This peak shift phenomenon is explained by the change in the heat transport (dashed curves) through the power generator shown in the figure as well.

5.4.4   *Entropy generation in power generation process*

We investigate the exergy flow in generic thermoelectric energy conversion systems, while the exergy [25] is a measure used for evaluating the quality of heat. Firstly, entropy generation in a thermoelectric power generation system is quantized. The entropy generation in the system can be described using the temperature at the nodes along the heat flow as

$$\dot{S}_{gen} = \frac{\partial S}{\partial t} + \left(\frac{q_h}{T_h} - \frac{q_h}{T_s}\right) + \left(\frac{q_c}{T_c} - \frac{q_h}{T_h}\right) + \left(\frac{q_c}{T_a} - \frac{q_c}{T_c}\right) > 0 \quad (5.48)$$

while power output $w = q_h - q_c$. The first right term of the above equation is equal to zero considering the steady-state behavior of the model. By substituting heat flow q_h and q_c by definition of thermal resistances, the entropy generation becomes

$$\dot{S}_{gen} = \frac{(T_c - T_a)}{\psi_c T_a} - \frac{(T_s - T_h)}{\psi_h T_s} \quad (5.49)$$

Figures 5.17 and 5.18 show the examples of entropy generation for a variety of the ZT value and the asymmetricity of thermal contacts. The difference between the figures is the heat source temperature and one is extremely high 30,000 K while the latter is 375 K. The data shows optimal design cases. It is clear that the smaller thermal resistance ratio ψ_c/ψ_h yields less entropy generation. It is explained by Fig. 5.16. Note that the entropy generation values are significantly different between the graphs. The heat flow at optimum design is

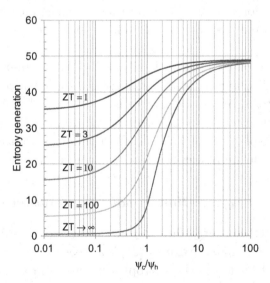

Fig. 5.17: Entropy generation as a function of the ratio of external thermal resistances for different ZT values. $T_a/T_s = 0.01$ for $T_a = 300$ K.

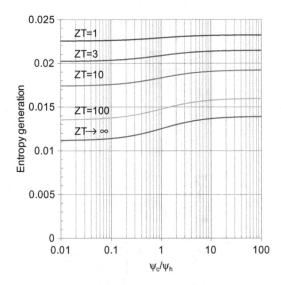

Fig. 5.18: Entropy generation as a function of the ratio of external thermal resistances for different ZT values. $T_a/T_s = 0.8$ for $T_a = 300$ K.

lower when ψ_c/ψ_h is smaller. This is essentially caused by the lower entropy generation.

It is obvious from Eq. (5.49) that the entropy generation linearly results in inverse of the sum of thermal resistances $(1/\sum \psi)$. The value of entropy generation changes by changing $\sum \psi$ but the above curves stay the same.

As Bejan [26] pointed out, the minimization of entropy generation is equivalent to the maximization of power output with symmetric dissipations. We reach the same conclusion only at infinite values of ZT, which occurs only in the ideal or reversible engine.

5.4.5 *Exergy flow in optimum system*

By definition, exergy is the metric measure of the maximum possible work in a process. In this particular thermoelectric system, the exergy per unit area Ξ [W/m^2] is found as the product of the heat flow per unit area in steady-state and Carnot efficiency. The entire system can be expressed as shown in Fig. 5.19, where destroyed exergy exits to the right and the delivered exergy exits downward.

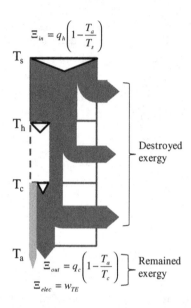

Fig. 5.19:   Exergy flow of a thermoelectric generation system.

The exergy is destroyed at the three stages of the system, including ψ_h, ψ_c, and thermoelement. Thus, remaining exergy Ξ is found as

$$\Xi = w + q_c \left(1 - \frac{T_a}{T_c} \right) \tag{5.50}$$

where w is the TE power output per unit area and q_c is the wasted heat per unit area flows in the cold side thermal path.

Figure 5.20 shows an example of the exergy per unit area at the optimum design with respect to ZT for different external thermal conditions, where $\psi_c + \psi_h = 1$ (constant). Electricity contribution of the exergy significantly increases by increasing ZT and gradually converges to a certain level. The electricity contribution is almost identical for any asymmetric thermal resistance systems. For a smaller ψ_c, remaining exergy, which is unconverted heat, is observed to be smaller and consequently the system generates the highest quality of energy output at the smallest ψ_c. As ψ_c/ψ_h increases, the remaining exergy increases. It is caused by the larger heat contribution of the cold side heat sink downstream of the system. This large amount

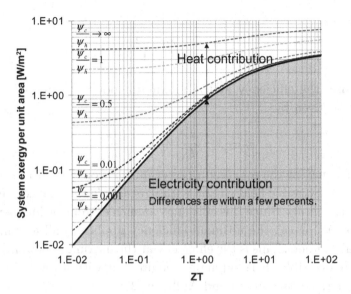

Fig. 5.20: The normalized exergy as a function of the ratio of external thermal resistances (ψ_c/ψ_h) and for different ZT values. $T_a/T_s = 0.8$ for $T_a = 300$ K.

of heat contribution for large ψ_c/ψ_h (> 1) suggests the potential opportunity of co-generation.

5.5 Power Generation Limit with Infinitely Large Z

Investigation on the limit of infinitely large Z provides a generic trend of the power output. Since the temperatures T_h and T_c have a weak dependency on (ψ_c/ψ_h) at the optimum power output design (electrothermal co-impedance matching), the efficiency changes slightly. The expression of the efficiency $\eta = W/Q$ is rewritten from Eq. (5.31) as

$$\eta = \left(1 - \frac{T_c}{T_h}\right)\left(\frac{m-1}{m + \frac{T_c}{T_h}}\right) \tag{5.51}$$

where the first part of the right-hand side is equal to the Carnot efficiency. This suggest that the efficiency at the maximum power output will converge to the Carnot efficiency as $Z \to$ infinity, where $m \to$ infinity as well. At the extreme, Eq. (5.29) converges to the following:

$$\frac{T_s}{T_a} \to \left(\frac{T_h}{T_c}\right)^2 \tag{5.52}$$

Therefore, the efficiency at the maximum power output when Z→ infinity is given by

$$\eta \to 1 - \frac{T_c}{T_h} = 1 - \sqrt{\frac{T_a}{T_s}} \tag{5.53}$$

The efficiency at the maximum power output converges to the exact same as a reversible heat engine with irreversible thermal contacts as Z goes to infinity. Since this efficiency at such extreme is independent of either ψ_c or ψ_h, the ultimate efficiency applies to all asymmetric thermal contacts. This is the efficiency at the maximum power output for an irreversible thermodynamic engine but with a reversible heat engine derived by Curzon and Ahlborn [27].

5.5.1   *Asymmetric contacts*

In order to investigate the behavior of thermoelectric generators with a very large Z, influence of the asymmetric thermal contacts need to be considered. Firstly, leg length of the thermoelectric element is varied to find its optimum value for maximum power output in a symmetric thermal contact case, while the electrical load resistance is always optimized. Then, the power output and efficiency are calculated with the same leg length for asymmetric thermal resistance cases, in which total thermal resistance $\psi_c + \psi_h$ is assumed to be constant. Figure 5.21(a) shows how the asymmetric thermal contact influences the optimum leg length. Figure 5.21(b) shows how the efficiency changes. Smaller cold side thermal resistance relative to the hot side always shows better efficiency and the larger one shows the opposite trend.

5.5.2   *Curzon–Ahlborn limit*

Figure 5.22 shows the higher bounds of the system efficiency at the maximum power output from the above cases when Z is very large. In the figure, the Carnot efficiency is defined, as it should, as a function

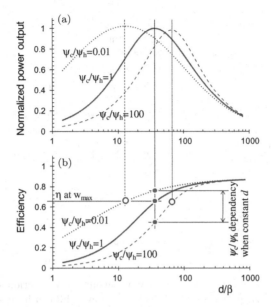

Fig. 5.21: Power output normalized to the maximum at $\psi_c/\psi_h = 1$ and efficiency as a function of normalized leg length. $T_a/T_s = 0.1$, $Z = 1$ (order of $ZT \sim 10^3$), $\psi_c/\psi_h = 0.01$, 1, and 100.

of reservoir temperatures $(1 - T_a/T_s)$. Interestingly, the asymmetric limits show a very different behavior with a low thermal resistance at either hot or cold side, when the leg length is fixed to the value equal to the optimum for the symmetric system. See the curves indicated by $\psi_c/\psi_h \to \infty$, with d_{opt} at $\psi_c/\psi_h = 1$ and $\psi_c/\psi_h \to 0$ with d_{opt} at $\psi_c/\psi_h = 1$. These limits are very similar to the ones for generic cyclic thermodynamic systems reported by Esposito *et al.* [28]. On the other hand, when the leg length is fully optimized, the thermoelectric generator recovers the Curzon−Ahlborn limit.

5.5.3 *Efficiency and power output*

Figure 5.23(a) shows the power output as a function of normalized efficiency with respect to the Carnot value when $T_a/T_s = 0.2$ as an example. The ZT value is modified by changing only the thermal conductivity. The leg length d is a variable along the curves. Curves start from the point all values are zero and increase in both

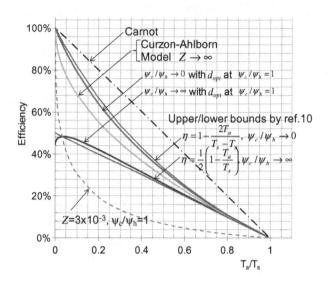

Fig. 5.22: Efficiency at maximum power output as a function of T_a/T_s, the model at infinite Z, perfectly matches the Curzon−Ahlborn limit at any ψ_c/ψ_h. The limits of the thermodynamic cyclic engines and the curves for $Z = 3 \times 10^{-3}$ with $\psi_c/\psi_h = 1$ are also shown.

Source: Reproduced with permission from *J. Appl. Phys.*, **111**, 024509 (2012).

power output and efficiency as leg length d increases. After reaching a peak, power decreases, but efficiency continues to increase. This trend is observed for any ZT value. Only for the case of $ZT \to$ infinity, the maximum efficiency exactly matches the Carnot efficiency, i.e. $(1 - T_a/T_s)$, where leg length d becomes extremely large and then power output diminishes. Figure 5.23(b) shows the normalized power output with different T_a/T_s ratios as the function of normalized efficiency. All the curves are calculated with the ZT of infinity. The outside of the hatched area on the left-hand side of the curves shows the region where no design case exists.

Carnot limit is always reached when the output power is zero. In the case when ambient reservoir temperature is much smaller than the heat source temperature $(T_a/T_s = 0.001$ in Fig. 5.23), the maximum output power is achieved at the point where efficiency comes close to the Carnot limit. The temperature gradient is so large that one can get lots of power with very high efficiency. For the opposite

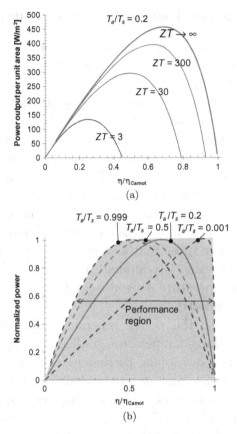

Fig. 5.23: Output power as a function of efficiency: (a) varying ZT for $T_a/T_s = 0.2$ and (b) varying T_a/T_s with power normalized with respect to the peak power value of the individual curves. $T_a = 300$ K fixed, $\sum \psi = 1.0$, $\psi_c = \psi_h$ and only thermal conductivity is modified for various ZT.

extreme case, when T_s is close to T_a, the maximum power is observed when the efficiency equals half of the Carnot efficiency.

Given any thermal contact resistance with reservoirs, the design of the thermoelectric leg thickness allows to achieve either the maximum power output or the maximum theoretical efficiency (while power equals zero). The electrical load resistance is always chosen to be optimum (electrically maximum). The peak value of the maximum power output and maximum theoretical efficiency depends on material increases with improving ZT.

5.6   Multi-Segment Thermoelectric Generators

Figure 5.24 shows the simplest configuration of the segmented thermoelectric leg in a module, which consists of two different materials
in each p-type or n-type leg.

By nature of thermodynamics, a large energy conversion efficiency
at maximum power output is expected as a large temperature gradient is applied. Under a utilization design of thermoelectric to a
large temperature difference between the hot and cold reservoirs, the
optimum design temperature T_h and T_c can be significantly different,
where the material properties may not well cover the entire temperature difference. Segmentation is a technique that connects two or
more different materials in a leg as an effort to boost the power output or efficiency over those that a single material can achieve. This
is because the ZT value is temperature-dependent (see Fig. 5.3 in
Section 5.2 for temperature dependency of the ZT values for SiGe
thermoelectric materials).

Segmentation for the power generator application is limited
because it requires large temperature gradients during operation conditions. In addition, developing thermomechanically stable contacts
between the materials often prove to be challenging. In space application, segmented structures have been investigated and optimized [30].
Full details of modeling and optimization for a variation of $p - n$
unicouples is found in Ref. [31]. Another optimization study investigated the so-called compatibility factor and observed a considerable
difference in efficiency in segmentation [32] with different compatibility factors. Even if the material ZT is maximized at each segment,

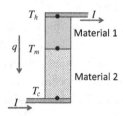

Fig. 5.24:   Basic configuration of a multi-segment thermoelectric leg. Two segments are shown as an example. This particular leg is n-type. Material 1 has a
peak ZT at higher temperature than Material 2.

the overall efficiency could be low if the compatibility factors of the materials from different segments are not sufficiently close to each other because the optimal operation of each segment cannot be met under a universal current flow all across the segments in this case. Therefore, both the ZT values and the compatibility factors must be taken into account when segmented legs are designed.

Mechanical compliance at the contact between segments is also indispensable [33]. In a space mission application, the radio isotope generator (RTG) was investigated at the Jet Propulsion Laboratory (JPL) in NASA focusing on segmented advanced skutterudite-based thermoelectric materials [34]. The inhomogeneous materials are brazed with an interfacial metal with a contact resistance of 15 $\mu\Omega\,cm^2$ or less and the unicouples are tested at hot junction up to 625°C and cold junction maintained to 200°C for about 10,000 hours.

5.7   Impact of Temperature-Dependent Properties

Thermoelectric material properties are often referred to by the figure of merit as the ZT value in non-dimensionalized form. The temperature (the T-part) in the definition of ZT is the mean temperature under the operational condition. The Z value is also heavily temperature-dependent. (see Section 5.2)

The simplest strategy is to find the peak temperature of the ZT value with the nominal temperature of the application. If the thermal resistances of the heat sinks are not known, a preliminary estimate of the nominal temperature can be the mean temperature of the entire temperature range, i.e. $\bar{T} \approx (T_s + T_a)/2$ between the hot reservoir and the cold one. A thermoelectric material, which has a peak ZT at this $\bar{T}$, is supposed to perform close to the optimal found for the application. The optimization of the module design may then be further conducted for fine-tuning. Considering other optimization options, the heat sinks of either the hot or cold side (or both) can be tuned to achieve the temperature peak of the specific material performance. In this process, to find the temperature match, temperature dependency of each of the three components of the Z value may not be fully considered. This is still fine as far as the thermoelectric module design is optimized for the maximum power output.

Thermodynamic limitation is another key consideration in addition to the temperature dependency of material properties. Energy conversion at the maximum power output condition is found as the maximum efficiency of the thermoelectric derived as the following if both side temperatures T_h and T_c are known:

$$\eta_{atW_{max}} = \left(1 - \frac{T_c}{T_h}\right)\left(\frac{\sqrt{1+ZT}-1}{\sqrt{1+ZT}+\frac{T_c}{T_h}}\right) \qquad (5.54)$$

The first part of the right-hand side of this equation is the Carnot efficiency and then the second part is a material property-dependent factor, which was discussed earlier in this chapter. The Carnot efficiency depicts the thermodynamic limitation based on the temperature range of the application, which is independent of the materials used. Figure 5.25 shows how the thermodynamic factor changes the power output and efficiency with varying source temperature and device figure of merit. In the figure, a Gaussian function of ZT with temperature shown in the left plot was assumed for a sample material, and then the maximum efficiency was calculated for this material as a function of source temperature in the right plot, while T_a is

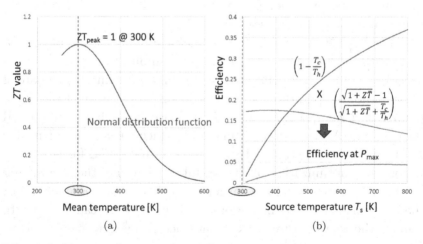

(a)   (b)

Fig. 5.25: Impacts of generic thermodynamic and material properties on the efficiency: (a) a sample Gaussian function of ZT with temperature used and (b) the corresponding maximum efficiency of the material as a function of the source temperature with breakdown of the thermodynamic limit (Carnot) and the material-dependent part of the efficiency.

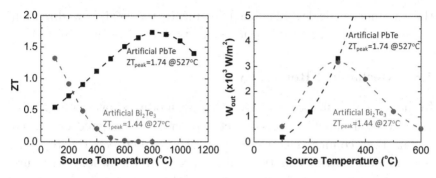

Fig. 5.26: Temperature profile of the artificially generated temperature-dependent material ZT and the power output with these materials.

fixed at 300 K. In this calculation, thermal resistances of the hot and cold sides are assumed to be the same (symmetric). Note that T_h and T_c are calculated from the reservoir temperatures to maximize the power output. Note that the temperature of peak output ($\sim$ 650 K) is significantly higher than the temperature at peak ZT (300 K) because of the high thermodynamic efficiency component at higher temperatures.

Figure 5.26 shows another example comparison between artificial examples of bismuth telluride (Bi_2Te_3) and led telluride (PbTe) materials. The left plot shows the temperature-dependent ZT values used for the two materials, and the right plot shows the maximum power output as a function of source temperature for the two materials. It is interesting to see that Bi_2Te_3 that has lower average ZT and lower peak-ZT temperature generates a larger power than PbTe below $T_s = 300°C$, which has rarely been considered as a workable temperature range for this material. This is because the ZT values at low-temperature region still largely influence the power output even if the source temperature is much higher. In order to address such an impact of temperature profiles of material properties, another figure of merit called the engineering figure of merit has been introduced to adequately quantify the realistic efficiency and power output under a large temperature range [35, 36].

Here, we see that the temperature dependency is more important for power generation typically with a large temperature difference, where the cooling applications have less impact. Segmentation of the thermoelectric leg (element) in line to the heat flow direction mitigates the temperature dependency problem (see Section 5.6).

5.8 Mechanical Reliability

In the typical Π-shaped structure of a TE module shown in Fig. 5.27, the electrical and thermal contacts are made of soldered or brazed metals that connect electrodes and TE legs between the top and bottom substrates. The two substrates hold the whole structure by mechanically and rigidly anchoring the surface microstructure. Hence, there are many layers involved in the structure, as shown in the right image of Fig. 5.27. The assembly of this complex multi-layer structure is placed under thermal gradients to harvest electricity out of the heat flow. In this process, thermal expansion coefficient (CTE) mismatch between the materials can turn into a large thermal stress at the junction due to a large temperature gradient across the structure. The mechanism of creating such a large thermal stress is not only from the CTE mismatch of materials but also from the elastic modulus variation across the entire structure. The elements located near the edge of the module tend to receive a much larger

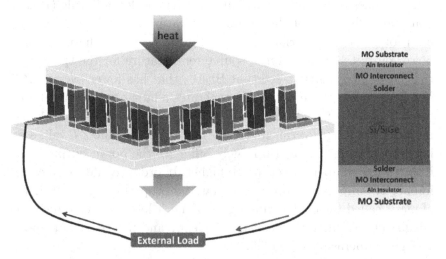

Fig. 5.27:   Π-shaped module structure and multi-layers consisting of a thermo-electric leg with Si/SiGe superlattice thermoelectric materials on a molybdenum substrate.

deformation than the center region because of the larger deformation of the substrates near the edge. If both of the substrates are made of the same material and same dimensions, the hot side substrate tends to expand more than the cold side substrate due to the difference of temperature change. In free suspension, the entire module tends to bend because of this non-uniform deformation. However, the substrate is not allowed to deform in the cross-plane direction due to the mechanical constraints by the physical contact of the heat source and heat sink or the mechanical fixture. Therefore, the substrate tends to expand mostly in the in-plane direction. This results in the significantly larger stress and thus larger deformation at the contacts of the corner legs.

In the following section, the thermal stress analysis is summarized. More details about the analysis is found in Ref. [37]. Here, we deal with a particular material configuration designed for a high-temperature application. As shown in Fig. 5.27, an aluminum nitride (AlN) insulator layer and a molybdenum (Mo) interconnect layer are firstly deposited on an Mo substrate. Then, the silicon and silicon germanium (Si/SiGe) superlattice legs are soldered with both substrates. There can be a stress created at the solder contact even when the module is not under temperature gradients, which is created by gradually cooling down from the soldering temperature to the room temperature. However, the solder contacts can absorb quite a lot of stress due to its viscoelastic characteristics. Hence, the residual stress in room temperature is considered to be negligible.

The maximum shear stress that is created at the solder contact during the power generation operation under ΔT is given by Ziabari *et al.* [37] as

$$\tau_{\max} = k\frac{\alpha\Delta T}{\lambda_1}\tanh kl$$

$$\times \left[1 + \frac{\left(\tanh kl + \coth kl + 8kl\left(\frac{L}{2l} - 1\right)\right)}{\left(\left(\tanh kl + \coth kl + 4kl\left(\frac{L}{2l} - 1\right)\right)^2 + 8kl\left(\frac{L}{2l} - 1\right)\tanh kl + 4kl\left(\frac{L}{2l} - 1\right)^2\right)\sinh 2kl}\right]$$

$$(5.55)$$

where l is the half-width of TE leg, L is the end-to-end distance between the two legs in a module, ΔT is the temperature difference between the hot and cold sides, k and λ_s are defined as follows:

$$k = \sqrt{\frac{2\lambda_s}{\kappa}} \qquad (5.56)$$

$$\lambda_s = \frac{1 - v_s}{E_s d_s} \qquad (5.57)$$

where k is the parameter of the interfacial shearing stress, κ is the total interfacial shear compliance of the mid-layers between the two top and bottom components, α [1/°C], v_s, E_s [GPa], and d_s [m] are the coefficient of thermal expansion (CTE), Poisson ratio, elastic modulus, and the thickness of the substrate, respectively.

Note that the above is a case of two-dimensional analysis, while the real module is three-dimensional even if the shape is symmetric. In order to find the quantitative stress, 3D numerical simulation needs to be conducted. Figure 5.28 shows several example cases of 2, 6, and 36 legs in a thermoelectric module with variations of fill factor F and mechanical boundary conditions.

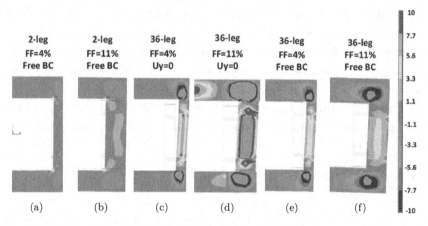

Fig. 5.28: Shear stress distribution in TE module structures, maximum and minimum are capped at ±10 MPa: (a) free-standing two-leg module with $F = 4\%$; (b) free-standing two-leg module with $F = 11\%$; (c) 36-leg module constrained at the cold side with $F = 4\%$; (d) 36-leg module constrained at the cold side with $F = 11\%$; (e) free-standing 36-leg module with $F = 4\%$ and (f) free-standing 36-leg module with $F = 11\%$.

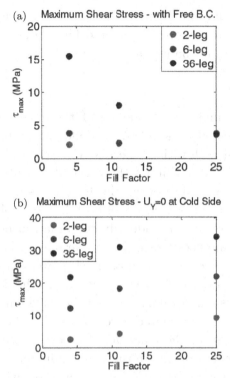

Fig. 5.29: Maximum shear stress at the interconnect/solder interface against fill factor for three different structures with (a) free-standing boundary condition and (b) perpendicular translation constrained at the bottom interface.

Figure 5.29 summarizes the maximum shear stress as a function of fill factor F [%] for different cases of 2-leg, 6-leg, and 36-leg structures and two different boundary conditions: (a) both sides free-standing and (b) cold side constrained. For free-standing boundary condition, the maximum shear stress increases with decreasing fill factor, except for the 2-leg module. For large $F \geq 25\%$, the maximum shear stress for the three structures is not very different from each other. This is mainly due to the asymmetric bending of the two large substrates. The outward/inward expansion of the hot and the cold substrates will deform the TE legs of the low fill factor structure and generate strong shear stresses.

When the legs are anchored to the bottom (limiting the perpendicular translation at the cold side), the same trend is observed for all the three modules in (b) with the stress increasing with increasing fill factor. A larger stress is created for a larger number of legs, reaching ~35 MPa for 36 legs at fill factor 25%.

5.9 Summary

In this chapter, analytical modeling of thermoelectric power generation was discussed in detail. A single-leg generator was modeled and analyzed first with various thermal and electrical parasitic losses considered. The model includes the thermal spreading between the leg and the substrate as well as the heat transfer across the heat sink to the temperature reservoirs of both hot and cold sides. Hence, this model helps to not only identify the practical performance but also analyze the impact of each design parameter. For example, the model can be used to examine how much power output and energy conversion efficiency will change in response to the design change of an external heat sink. Optimization for the maximum power output by minimum use of the material resources was then modeled and analyzed. The mass of the material is directly reflected in the cost analysis of the generator, hence the cost minimum optimization was discussed. Integration of thermoelectric generator with heat sinks or microchannels is quite a reasonable approach due to the flat shape of the thermoelectric modules. Since such heat sink requires a certain level of pump work, energy payback was discussed. As the heat flux increases across the thermoelectric, the pump work increases significantly over the heat flux of 1×10^5 W/m^2, until it reaches to the point that payback power for the pump and fan is reasonably small.

There are several different technical aspects needed to design a reliable and high-performance thermoelectric power generator. The analysis with large ZT value showed the direction and upper limits of the thermoelectric power generators. Materials have attracted the most attention. It is important to have a wide spectrum of the temperature-dependent $ZT(T)$ in addition to improve the peak ZT value. For the wide temperature difference application, such as the fuel-burned system, multi-segmented (leg) modules become the key enabler. Thin film regime is necessary for high heat flux integration.

Thermomechanical analysis showed the critical stress concentration at the contacts of thermoelectric legs.

References

[1] Laird Technology, Available at: https://www.lairdthermal.com/.

[2] F. D. Rosi, E. F. Hockings and N. E. Lindenblad, *RCA Rev.*, **22**, 82–121 (1961).

[3] D. Vashaee and A. Shakouri, Thermionic power generation at high temperatures using Si Ge/Si superlattices, *J. Appl. Phys.*, **101**(5), 053719 (2007).

[4] M. S. Dresselhaus, G. Chen, M. Y. Tang, R. G. Yang, H. Lee, D. Z. Wang, Z. F. Ren, J-P. Fleurial, and P. Gogna, New directions for low-dimensional thermoelectric materials, *Adv. Mat.* **19**(8), 1043–1053 (2007).

[5] Y. Pei, H. Wang and G. Jeffrey Snyder, Band engineering of thermoelectric materials, *Adv. Mat.*, **24**(46), 6125–6135 (2012).

[6] A. Majumdar, Thermoelectricity in semiconductor nanostructures, *Science*, **303**(5659), 777–778 (2004).

[7] J.-H. Bahk, Z. Bian and A. Shakouri, Electron energy filtering by a non-planar barrier to enhance the thermoelectric power factor in bulk materials, *Phys. Rev. B*, **87**, 075204 (2013).

[8] A. Shakouri, Recent developments in semiconductor thermoelectric physics and materials, *Ann. Rev. Mater. Res.*, **41**, 399–431 (2011).

[9] K. Yazawa and A. Shakouri, Proc. IMECE2010, *ASME* (2010).

[10] P. M. Mayer and R. J. Ram, *Nanosc. Microsc. Thermophys. Eng.*, **10**, 143–155 (2006).

[11] J. W. Stevens, *Energy Convers. Mgmt.*, **42**, 709–720 (2001).

[12] G. J. Snyder, Energy Harvesting Technologies, S. Priya, D. J. Inman (Eds.), Springer, New York, 330–331 (2009).

[13] M. Esposito, K. Lindenberg and C. Van DenBroeck, *J. Explor. Front. Phys.*, **85** 60010 (2009).

[14] X. W. Wang, H. Lee, Y. C. Lan, G. H. Zhu, G. Joshi, D. Z. Wang, J. Yang, A. J. Muto, M. Y. Tang, J. Klatsky, S. Song, M. S. Dresselhaus, G. Chen and Z. F. Ren, *Appl. Phys. Lett.*, **93**, 193121 (2008).

[15] G. Joshi, H. Lee, Y. Lan, X. Wang, G. Zhu, D. Wang, R. W. Gould, D. C. Cuffand M. Y. Tang, Enhanced thermoelectric figure-of-merit in nanostructured p-type silicon germanium bulk alloys, *Nan. Lett.*, **8**(12) 4670–4674 (2008).

[16] J.-H. Bahk, M. Youngs, K. Yazawa, A. Shakouri, In: *Proceedings of the 43rd ASEE/IEEE Frontiers in Education Conference* (2013).

[17] H. J. Goldsmid, Chapter 3 Conversion efficiency and figure-of-merit, D. M. Rowe (Ed.), *CRC Handbook of Thermoelectrics*, 19–25, CRC Press (1995).

[18] M. Freunek, M. Muller, T. Ungan, W. Walker and L. M. Reindl, *J. Elect. Mat.*, **38**(7), 1214–1220 (2009).

[19] S. Song, S. Lee and V. Au, Closed-form equation for thermal constriction/spreading resistances with variable resistance boundary condition, *Proc. of the 1994 International Electronics Packaging Conference*, 111–121 (1994).

[20] K. Yazawa and A. Shakouri, Cost-efficiency trade-off and the design of thermoelectric power generators, *Environ. Sci. Technol.*, **45**(17), 7548–7553 (2011).

[21] D. Kraemer, B. Poudel, H.-P. Feng, J. C. Caylor, B. Yu, X. Yan, Y. Ma, X. Wang, D. Wang, A. Muto, K. McEnaney, M. Chiesa, Z. Ren and G. Chen, High-performance flat-panel solar thermoelectric generators with high thermal concentration, *Nat. Mat.*, **10**(7), 532–538 (2011).

[22] J. A. Potkay, G. R. Lambertus, R. D. Sacks and K. D. Wise, A Low-power pressure- and temperature-programmable micro gas chromatography column, *J. Micro. Syst.*, **16**(5) (2007).

[23] K. Yazawa, G. Solbrekken and A. Bar-Cohen, Thermoelectric-powered convective cooling of microprocessors, *IEEE Trans. Adv. Packa. Technol.*, **28**(2), 231–239 (2005).

[24] K. K. Shah and M. S. Bhatti, Laminar convection heat transfer in ducts [Chapter 3], In *Handbook of Single Phase Convective Heat Transfer* (1987).

[25] K. Yazawa and A. Shakouri, Exergy analysis and entropy generation minimization of thermoelectric waste heat recovery for electronics, *InterPACK*, 2011–52191 (2011).

[26] A. Bejan, *Entropy Generation Minimization*, CRC Press, 249–252 (1948).

[27] F. Curzon and B. Ahlborn, *Am. J. Phys.*, **43**, 22 (1975).

[28] M. Esposito, R. Kawai, K. Lindenberg and C. Van Den Broeck, *Phys. Rev. Lett.*, **105**, 150603 (2010).

[29] K. Yazawa and A. Shakouri, Optimization of power and efficiency of thermoelectric devices with asymmetric thermal contacts, *J. Appl. Phys.*, **111**, 024509 (2012).

[30] M. S. El-Genk, H. H. Saber and T. Caillat, Efficient segmented thermoelectric unicouples for space power applications, *Energ. Convers. Manage.*, **44**(11), 1755–1772 (2003).

[31] M. S. El-Genk and H. H. Saber, Modeling and optimization of segmented thermoelectric generators for terrestrial and space applications, D. M. Row (Ed.), in *Thermoelectrics Handbook Macro to Nano*, CRC Press (2006).

[32] M. Zare, H. Ramin, S. Naemi and R. Hosseini, Exact optimum design of segmented thermoelectric generators, *Internation. J. Chem. Eng.*, (2016).

[33] S. A. Firdosy, B. C. Y. Li, V. A. Ravi, J. P. Fleurial, T. Caillat and H. Anjunyan, U.S. patent No. 9722163, U.S. patent and trademark office (2017).

[34] I. S. Chi, K. Smith, C. K. Huang, S. Firdosy, K. Yu, K. B. Phan and S. Chanakian, Advanced skutterudite-based unicouples for a proposed enhanced multi-mission radioisotope thermoelectric generator: An update, In *Meeting Abstract No. 27*, 1175–1175, The Electrochemial Society, (2017).

[35] Y. R. Koh, K. Yazawa, A. Shakouri, T. Nagahama, S. Maeda, T. Isaji and Y. Kasai, Analytical optimization of the design of film-laminated thermo-electric power generators, *J. Electron. Mater.*, **48**(11), 7312–7319 (2019).

[36] H. S. Kim, W. Liu, G. Chen, C.-W. Chu and Z. Ren, Relationship between thermoelectric figure of merit and energy conversion efficiency, *PNAS* **112** (27), 8205–8210 (2015).

[37] A. Ziabari, K. Yazawa and A. Shakouri, Designing a mechanically robust thermoelectric module for high temperature application, *Proceedings of International Thermal Conductivity Conference (ITCC) and The International Thermal Expansion Symposium (ITES)*, DOI: 10.5703/1288284315556, (2015).

Chapter 6

Industrial and Energy Applications of Thermoelectric Generation

This chapter highlights the several modeling and analysis cases of thermoelectric (TE) power generation for industrial and energy applications. The focused topics include the thermoelectric topping cycle for power plants and the high-temperature waste heat recovery from grass manufacturing processes. For the middle-temperature applications, the vehicle exhaust gas heat recovery and the solar concentrate combined heat and power application are also discussed. Solar CHP includes the results from a hands-on experimental demonstration. This chapter not only describes the examples but also shows how the analytic model and the optimization of thermoelectric generators works in the applications.

6.1 Power Plant and Topping Cycle

Steam turbine (ST) cycles are typically used in a power plant. In the cycle, there is a large temperature difference between the potential fuel combustion temperature up to around 2250 K (adiabatic) and the high-pressure steam temperature limited to operate at around 900 K. This temperature gap results in a large amount of thermodynamic losses during the cycle. A solid-state thermoelectric generator placed on top of the ST cycle can produce additional electrical power by utilizing the large temperature difference. By selecting the right materials for the TE generator for high-temperature operation, the energy production can be increased with the same fuel consumption.

Recent advances in nanostructured thermoelectric materials could provide practical performance benefits. Here, we present a theoretical study on the optimization of the interface temperature connecting these two idealized engines for energy economy as a combined system. We also analytically study the optimum point of operation between the maximum power output for minimizing the payback and the maximum fuel efficiency for the economy for each generator. The economic optimum may result in a significant reduction in energy cost ($/kWh) to produce electricity as the power plant. The combined TE topping generator system provides a lower energy cost for any period of operational life and lower interface temperature compared to the ST cycle alone. The maximum power output is observed at around 700 K of interface temperature for 10,000 hours of operation, which suggests a significant improvement for the turbine reliability. On the other hand, the minimum cost power production from the combined system is observed at the interface temperature at the upper limit (900 K) with the thermoelectric generator using an off-the-shelf material ($ZT = 1$).

6.1.1 *Energy cost and fuel consumption*

The tragedy in year 2011 at the nuclear plant in Fukushima raised public concerns about the safety of nuclear power not only in Japan but also throughout the world [1]. The national-level strategy for power plants may have to consider many additional factors. One of the factors is the energy price ($/kWh). Considering sustainability of oil resources and global warming, without a nuclear approach, we have no choice but to increase fuel efficiencies or explore renewable sources, such as solar and wind. The U.S. energy flowchart [2] shows that the energy actually used is only ~40% of the energy input. This study shows the importance of the approaches for increasing fuel efficiency in power production and lowering the energy cost ($/kWh).

Among various power generators available, STs, invented by Parsons [3] in 1884, have a long history. This particular mechanism is widely used for power generation with a simple system structure. STs are also one of the most efficient mechanical engines [4] as its intrinsic energy conversion efficiency is closer to the Carnot efficiency.

In addition to using saturated steam from a boiler, superheated steam is also used with a compressor for higher efficiency. The thermodynamic conversion efficiency, however, is limited by the inlet steam temperature since the structural material of turbines, such as stainless steel, has a temperature-dependent mechanical yield limit at points of extremely high steam pressure. For example, the theoretical upper limit of thermodynamic efficiency (Carnot efficiency) for a 900 K steam temperature and 300 K ambient temperature is 67%. Furthermore, since the system includes irreversible thermal contacts, one cannot obtain any usable power at the maximum efficiency. Theoretical efficiency of an ideal thermodynamic engine (with no intrinsic losses) with irreversible contacts is only 42% at the maximum power output based on the Curzon–Ahlborn limit [5]. In this section, validity of adding a TE generator on top of an ST cycle is investigated. The best operating point for fuel economy for the TE topping ST cycles will be found in between the conditions for the maximum efficiency and the efficiency at maximum power output. As an immediate summary, the best achieved efficiencies of other technologies as a function of heat source temperature are given by Vining [6].

6.1.2 *Topping cycle on ST*

The combined system we discuss is composed of a TE generator on top of a ST cycle, the most typical kind of Rankine cycle. The TE generates an additional amount of power by using the large temperature gap between the source temperature and the steam temperature. Even with a special design dedicated for high-temperature operation for turbines [7], the steam temperature is limited to $< 650°C$ (923 K). Some of the ST cycles work under saturated steam. For higher temperature input, recent STs use superheated steam. A Hirn cycle may be used as a similar cycle using a compressor to obtain a superheated steam under high pressure before boiling, while a Rankine cycle uses a pump to circulate the condensed water back into the boiler. In this analysis, however, we simply use the ST cycle as an irreversible engine since the primary interest is on the impact resulting from adding the TE generator on top. Then the interfacial temperature is optimized for the lowest energy cost.

TE generators have been receiving more attention for waste heat recovery applications, for which the temperature range is similar to that of saturated STs. There are several studies of TE generators and waste heat recovery systems. Chen *et al.* [8] reported a comprehensive review of the various applications. Chen *et al.* [9] studied two-stage thermoelectric generators. Caillat *et al.* [10] studied the graded multiple TE materials for maximizing the practical efficiency of power generation. Qiu and Hayden [11] studied a combined TE and organic Rankine cycles (ORC) to effectively use the heat exhausted from an organic Rankine cycle turbine with three different modes of TE generators. Gou *et al.* [12] studied the low-temperature waste heat recovery with TE.

Unfortunately, thermoelectric device in general provides only a moderate efficiency despite a lot of research efforts spent on the material science. The advantage of the thermoelectric is in the space-limited or weight-limited applications, such as the area of vehicle exhaust heat recovery [13–15]. Such utilizations have been heavily investigated. Thermoelectric energy conversion is theoretically scalable to temperatures much higher than superheated steam and TE materials can be selected to match the targeted operation temperature. This scalable characteristic becomes an advantage which other technologies cannot accommodate, where the temperature range of working fluid is limited. Also, several other aspects of high-temperature thermoelectric power generation have been investigated, e.g. concentrated solar TE generators by Kraemer *et al.* [16], Yazawa *et al.* [17], and Xiao *et al.* [18].

The TE module can be designed for optimum performance by changing the element thickness to match the external thermal resistances as reported in Ref. [19]. Using heat concentration by making the fractional cross-section area of the element smaller, only orders of magnitude smaller mass of the material is needed for the same power output [20] as far as the thermal resistance match is maintained. The thermal and electrical parasitic impacts are also investigated in Ref. [21] for the optimal design. This optimum TE design takes into account both the initial cost and the operating cost of the fuel.

Knowles and Lee [22] studied a combined system placing a TE on top of a Brayton cycle and reported the efficiency at the maximum power output. They pointed out that adding TE only adds the power at the lower temperature range of the turbine, but the efficiency cannot exceed that of the high-temperature gas turbine. They suggest that using a TE topping cycle is limited to cases for which space or price cannot be justified for a high-temperature turbine.

6.1.3 *Model for topping cycle*

The thermal circuit model of the system is shown in Fig. 6.1. T_g is the interconnecting temperature between the two engines. The power output from the TE is a function of the design parameter d, the thickness of the thermoelement. The parameters are always per unit area for this analysis.

The external thermal resistances ψ_h and ψ_c [K m^2/W] are assumed to be symmetric. This simplifies the equation since the

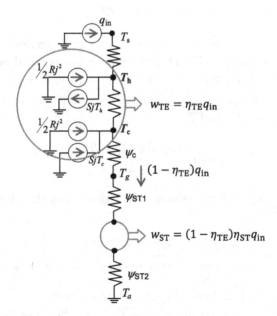

Fig. 6.1: Thermal network model of the entire system.

power output is insensitive to the asymmetry of the thermal contacts, where $\psi_h = \psi_c$. Thus,

$$\frac{(T_h - T_c)}{(T_s - T_g)} = \frac{d}{d + m\beta \sum \psi} = \frac{d}{d + d_0} \tag{6.1}$$

where d is the element thickness, d_0 is the optimum thickness for the maximum power output. β is the thermal conductivity of the TE material, and m is the ratio of the electrical load resistance against the internal resistance. T_s is the adiabatic flame temperature of the fuel, T_h and T_c are the hot side and the cold side temperature of the thermoelement, respectively. $\sum \psi$ is the sum of the external thermal resistances, thus $\sum \psi = \psi_h + \psi_c$. The optimum thickness for the maximum power output is found as

$$d_0 = m\beta \sum \psi \tag{6.2}$$

and

$$m = \sqrt{1 + Z\bar{T}} \tag{6.3}$$

where $Z\bar{T}$ is the dimensionless figure of merit of thermoelectric material and Z is defined as $Z = \sigma S^2 / \beta$, while σ is electrical conductivity and S is Seebeck coefficient. $\bar{T}$ is the mean operating temperature across the thermoelectric element. The generic power output w is found for a given T_h and T_c as

$$w_{TE} = \frac{m\beta Z (T_h - T_c)^2}{(1 + m)^2 d} \tag{6.4}$$

The efficiency η_{TE} of the power generation from the thermoelectric device is found as

$$\eta_{TE} = \frac{w_{TE}}{q_{in}} = \frac{m\beta Z (T_h - T_c)^2}{(1 + m)^2 d} \frac{1}{(T_s - T_h)/\psi_h} \tag{6.5}$$

where q_{in} is heat flux from the heat source flame.

While $\eta_{TE} \ll 1$ considering $ZT \sim 1$, the heat flow through the hot side and the cold side thermal resistances are similar. Thus, $(T_s - T_h)$ and $(T_c - T_g)$ are approximately equal. Based on this, the power output Eq. (6.4) can be rewritten as

$$w_{TE} = \frac{Z}{(1+m)^2} \frac{d/d_0}{(d/d_0 + 1)^2} \frac{1}{\sum \psi} (T_s - T_g)^2 \qquad (6.6)$$

$$\eta_{TE} \cong \frac{Z}{(1+m)^2} \frac{d/d_0}{(d/d_0 + 1)} (T_s - T_g) \qquad (6.7)$$

Here, power output is found as a function of the dimensionless thickness d/d_0. Other parameters are the given conditions, $\sum \psi$, T_s, and T_g. Both the extremely thin and thick thermoelement generate nearly zero output according to Eq. (6.7). This supports the assumption that an optimum exists.

Although the real material properties and Z value are temperature-dependent, they are kept constant across the temperature range. This is because, once the optimum temperature for TE is found, one can choose or develop the material match for the required temperature. Most of the currently available materials show $ZT \sim 1$ in average for a wide temperature range, as seen in Ref. [23]. Despite the fact that ZT is a temperature-dependent factor, we keep the properties constant so as to yield $ZT = 1$ at the interface temperature $T_g = 800$ K.

6.1.4 *Energy cost modeling*

Energy cost Y_{TE} [\$/kWh] for the thermoelectric section is calculated as follows:

$$Y_{TE} = \frac{I_{TE}}{H w_{TE}} + \frac{Y_f}{\eta_{TE}} \qquad (6.8)$$

where H is the operational life in hours for the design, Y_f is the cost of potential chemical energy of the fuel in units of \$/kWh, and I_{TE} [\$/m^2] is the initial material cost to build the system per unit area given by

$$I_{TE} = \rho d F G + 2\rho_s d_s G_s \qquad (6.9)$$

where subscript s indicates the substrates for the thermoelectric module, ρ is density, F is fill factor, and G is the material market unit price [\$/kg]. Heat sinks are not included in this part of the model, since they are already included in the Rankine cycle unit. We may replace them with modified versions but do not expect a different price, thus we did not double count the heat sink in Eq. (6.9).

The first term in Eq. (6.8) is the payoff for the initial investment. Thus, longer term operation yields a smaller energy cost since the initial investment cost is amortized over the number of operating hours. The second term is the operating cost, which is dependent on the fuel cost and the energy conversion efficiency. Therefore, maximizing efficiency or maximizing power impacts each individual term. As Eq. (6.6) for power and (6.7) for efficiency show, these factors are tightly related and therefore the relation yields a trade-off.

Substituting Eqs. (6.6), (6.7), and (6.9) into Eq. (6.8), the energy cost is found as a function of TE leg thickness, d:

$$Y_{TE} = \frac{(\rho d F G + 2\rho_s d_s G_s)(1 + m)^2 (d/d_0 + 1)^2}{HZ(d/d_0)(T_s - T_g)^2}$$

$$\times \sum_{TE} \psi + \frac{(1 + m)^2 (d/d_0 + 1)}{Z(d/d_0)(T_s - T_g)} Y_f \qquad (6.10)$$

For further simplification, we assume the ratio of the element thickness and the substrate thickness d/d_s is a constant. Considering the mechanical stiffness of the substrate and the necessity of heat spreading in the substrate, it is reasonable to assume a constant relation. Replacing d/d_0 with x, the above equation becomes

$$Y_{TE} = \frac{(\rho F G + 2\rho_s c G_s)(1 + m)^2 d_0 (x + 1)^2}{HZ(T_s - T_g)^2}$$

$$\times \sum_{TE} \psi + \frac{(1 + m)^2 (x + 1)}{Zx(T_s - T_g)} Y_f \qquad (6.11)$$

6.1.5 *Optimization of thermoelectric*

By taking the derivative of Eq. (6.11) to be zero, the value of x is found to minimize Y_{TE}. The solution is found as the following, which is the only real formula among the possible three solutions:

$$x = \frac{1}{3}\left(X + \frac{1}{X} - 1\right) \tag{6.12}$$

while

$$X = \frac{\sqrt[3]{2}A}{\sqrt[3]{-2A^3 + 27A^2B + 3\sqrt{3}\sqrt{27A^4B^2 - 4A^5B}}} \tag{6.13}$$

with

$$A = 2\frac{(\rho FG + 2\rho_s cG_s)(1+m)^2 d_0}{HZ(T_s - T_g)^2}\sum_{TE}\psi, \text{ and } B = \frac{(1+m)^2}{Z(T_s - T_g)}Y_f \tag{6.14}$$

6.1.6 Optimization of Rankine cycle

The ST can be considered with a simpler model, which comprises an ideal engine with irreversible thermal contacts for both hot and cold sides. The work generated by the turbine is considered as electrical power. The efficiency of the ST cycle η_{ST}' includes all of the mechanical energy transfer efficiency, $\sim 65\%$, combining the turbine and the compressor/pump and the mechanical-to-electrical conversion efficiency, $\sim 92\%$, out of the thermodynamic efficiency η_{ST}. If the cycle is ideally flexible for the 'design to the target', the efficiency is considered as $\eta_{ST} = (1 - T_a/T_g)$ for the reversible core of the engine with a coefficient C_{ST}. Thus, the efficiency of the ST cycle containing irreversible contacts becomes $\eta_{ST}'/C_{ST} = \eta_{ST}$:

$$Y_{ST} = \frac{I_{ST}}{Hw_{ST}} + \frac{Y_f}{C_{ST}\eta_{ST}} \tag{6.15}$$

From the conservation of energy and constant entropy, the relation of the power output and the efficiency is given by

$$w_{ST} = \frac{1}{\sum_{ST}\psi}\frac{\eta_{ST}'/C_{ST}(T_g - T_a - T_g\eta_{ST}'/C_{ST})}{1 - \eta_{ST}'/C_{ST}} \tag{6.16}$$

The initial cost I_{ST} is considered to build the system for maximum power. By following the Curzon–Ahlborn limit, maximum power is given by the contact temperatures T_g and T_a:

$$I_{ST} = U_{ST} w_{ST\_max} = U_{ST} \frac{1}{\sum_{ST} \psi} \left(T_g - 2\sqrt{T_a T_g} + T_a \right) \quad (6.17)$$

where $\sum_{ST} \psi$ is the sum of the irreversible thermal resistance per unit area $[\text{K m}^2/\text{W}]$ within the ST. Substituting Eqs. (6.16) and (6.17) into Eq. (6.15),

$$Y_{ST} = E \frac{(A - \eta_{ST}')}{(B - C\eta_{ST}') \eta_{ST}'} + \frac{D}{\eta_{ST}'} \quad (6.18)$$

where

$$A = C_{ST}, \ B = (T_g - T_a) C_{ST}, \ C = T_g, \ D = Y_f,$$
$$E = \frac{Y_f}{H} \left(T_g - 2\sqrt{T_a T_g} + T_a \right) C_{ST} \quad (6.19)$$

Similar to the thermoelectric engine, we find the efficiency. The solution is found as

$$\eta_{ST}' = \frac{(AE + BD) - \sqrt{(AE + BD)^2 - B(CD + E)(AE + BD)/C}}{(CD + E)} \quad (6.20)$$

Due to the model simplification, the components that are not considered here are as follows: (1) the efficiency degradation through the auxiliary components, e.g. pressure losses of working fluid, (2) temperature dependency of thermal properties of the working fluid, and (3) nonlinear mechanical losses of turbines and generators. However, in this analysis, the impact of interfacial temperature is the primary interest and the difference between a single ST and the TE combined system. To evaluate the differences, using the theoretical upper limit of ST would be reasonable to measure the validity of adding TE. Instead of considering a particular system, this model provides seamless scalability for the interfacial temperature.

6.1.7 Impact of ZT value

Returning to the TE generator, it is not obvious from Eq. (6.11), but careful investigation reveals that a material with a larger figure of merit could provide a lower energy cost. The infinitely large ZT gives $Y_{TE} \rightarrow Y_{TE.ZT=1}/(3+2\sqrt{2})$ at the maximum power output, while the load resistance ratio $m = \sqrt{1 + ZT}$. Thus, improving the material figure of merit (ZT value) is an important technological development. Thermal conductivity is the most influential parameter for the lower mass use of these expensive materials.

Shakouri [24] provided a comprehensive summary of recent developments. Due to the nature of thermoelectric properties, it is quite difficult to find a natural material which has a very large ZT. Similar to high-speed electronic applications where electrical properties of semiconductors are engineered, thermoelectric materials are also engineered by manipulating the atomic scale characteristics. It is well known that the thermal conductivity of semiconductors consists of two components. One is lattice thermal conductivity which is due to the phonon heat transport and the other is electronic thermal conductivity. The latter is constrained to electrical conductivity by the Weidman–Franz law. Phonon scattering approach [25] to reduce the lattice thermal conductivity is performed by superlattices, embedded nanoparticles, and rough nanowire structures. Based on the *Handbook of Thermoelectrics*, some semiconductors are available to work at high temperatures such as silicon–silicon germanium (Si/SiGe) used in radioisotope thermoelectric generators with a maximum ZT at 900 K or higher. Also, Boron compounds can work at 1300–1500 K. This is one of the advantages of solid-state thermoelectric devices for applications similar to this study.

6.1.8 Parametric case study

In the following, a parametric case is examined about the impact of the design factors, which are the leg length for the thermoelectric topping cycle and the standalone efficiency of the bottoming cycle. Then, integrated system performance is investigated.

As a base line for the thermoelectric module, the material ZT value is approximately unity, with the constant properties: thermal conductivity $\beta = 1.5$ W/(m K), electrical conductivity $\sigma = 25000$ S/m, Seebeck coefficient $S = 2 \times 10^{-4}$ V/K, density $\rho = 8200$ kg/m^3, and fill factor $= 10\%$ (fractional area coverage of TE element relative to the cross-section area of the heat flow).

The raw thermoelectric material price is 500 \$/kg. Thermal resistances of contacts ψ_h and ψ_c are 0.005 [K m^2/W] each for thermoelectric heat sinks. Thermal resistances in ST $\psi_{ST1} = \psi_{ST2}$ are pseudo-defined [K m^2/W]. Flame temperature is $T_s = 2250$ K and the ambient $T_a = 300$ K.

ST generator's performance constant C_{ST} is assumed to be $65\% \times 92\% = 60\%$, where the shaft power output is converted to electricity with an efficiency of 92%. The machine cost of the bottoming cycle is rarely found. For this case, the ST system machine cost (overnight cost) is set to 7000 \$/kW. Fuel cost is 0.108 \$/kWh as the primary energy cost per chemical energy potential in kWh. The calorific value of fuel comes from Ref. [26]. To make the cost impact clear, the auxiliary costs or interposed costs are not considered for the calculation. Interestingly, the price for calorific value was quite similar for gasoline and liquid propane gas (LPG) as of year 2015. This cost value is also the bottom line for the electricity supply cost.

The adiabatic flame temperature is from Ref. [27]. The analysis is carried out as a function of the operational life in hours for each design h and the interface temperature T_g.

6.1.8.1   *Thermoelectric generator part*

Figure 6.2 shows (a) the power output and heat flux per unit cross-section area of heat flow and (b) the energy conversion efficiency. Both are functions of the dimensionless leg thickness, d/d_0. The maximum power output at time equal to zero is found at $d = d_0$. A thicker TE element leads to a larger efficiency, but it significantly limits the heat flow, thus the power output decreases as the element becomes thicker.

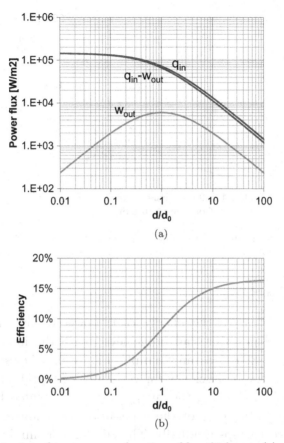

Fig. 6.2:   Output performance as a function of leg thickness: (a) power output
and heat flow per unit area and (b) energy conversion efficiency.

Figure 6.3 shows the dimensionless thickness (d/d_0) optimized
for the lowest energy cost and energy conversion efficiency both as
functions of the operation hours. Here, d_0 is the leg thickness optimum for maximum power output, which does not change by increasing operation hours with ideal material without time-dependent ZT
degradation. Over 2,000 hours, the optimum thickness is larger than
that of the maximum power output. This is due to better fuel efficiency. As the operation hours become longer, the economical optimum design approaches the maximum efficiency.

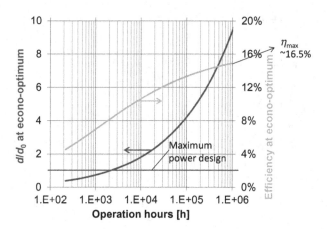

Fig. 6.3: Econo-optimum normalized leg thickness d/d_0 and the efficiency vs hours of operation.

6.1.8.2 *ST part*

Figures 6.4–6.6 show the energy cost, power output, and econo-optimum efficiency that can be achieved with the interfacial temperature $T_g = 800$ K as an example. Figure 6.4 shows the energy cost as a function of efficiency for variations of operation hours. For a small number of hours, the energy cost is dominated by the payback for the initial cost of the generator. For a longer operation, the energy cost drops gradually. The efficiency at minimum energy cost depends on the operating hours. This cost-minimum efficiency is found to be lower for longer operational life. Finally, for very long operating hours, the energy cost continues to drop as the efficiency increases.

The relationship between the power output and the efficiency of the idealized thermodynamic cycle is shown in Fig. 6.5. Any thermodynamic system which has irreversible thermal contacts exhibits similar behavior. The power output increases with increasing internal thermal resistance by increasing temperature difference across the generator. At some point, the power output will reach the maximum and then decrease while efficiency keeps increasing since the temperature difference is still increasing. Finally, the infinitely large internal resistance yields the maximum efficiency, but the power goes

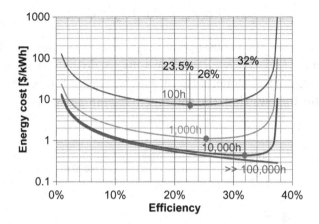

Fig. 6.4: Energy cost vs operating efficiency for Rankine cycle. $T_g = 800$ K.

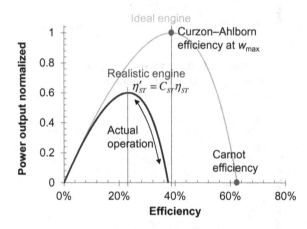

Fig. 6.5: Power output vs operating efficiency at $T_g = 800$ K while $T_a = 300$ K. The shrunk curve includes the efficiency coefficient $C_{ST} = 60\%$ relative to the Carnot engine. The economic optimum is found between the conditions for maximum power output and maximum efficiency.

to zero since the heat can no longer be transferred. In this analysis, the real maximum efficiency of the ST is determined by the performance factor C_{ST} relative to the Carnot efficiency. Actual operating points slightly deviate from the maximum power output for higher efficiency to obtain a better fuel economy.

6.1.8.3   *Power output of the combined system*

Figure 6.6 shows the econo-optimum power output for the TE part, the ST part, and the combined system. A dotted curve shows the econo-optimum ST alone if there was no upper temperature limit. Figure 6.6(a) shows that the combined system generates more power

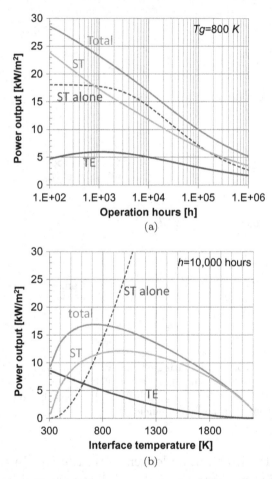

Fig. 6.6:   Economical optimum power output for the lowest energy cost as (a) functions of operation hours and (b) varying interface temperature for a case of 10,000 hours of operation. 'ST alone' refers to ST only while 'TE' and 'ST' indicate the TE generator and the ST cycle components of the combined system. ZT is approximately 1.

compared to the bottoming cycle (ST) alone from the same heat source. The output power tends to decrease due to the econo-optimum shift to the design for efficiency at longer life operation. We observe a peak for TE at around 1,000 hours. Figure 6.6(b) shows the impact of the interface temperature T_g on power output at an operation life of 10,000 hours. We observe an optimum T_g for maximum power output at around 700 K. The temperature at which the power output peaks for the combined system is lower than that of the bottoming cycle ($\sim$ 1100 K). This is due to the continuously decreasing output from TE as the interface temperature increases.

6.1.8.4 *Efficiency of the combined system*

Figure 6.7 shows the efficiency calculated for the same systems as above. For the design for longer operation hours, both TE and ST converge to the maximum efficiency for each individual system. The figure also includes the impact of enhancing the material ZT values of 1, 2, and 5 by changing thermal conductivity. The $ZT = 2$ curve corresponds to a range of advanced high-end materials ever fabricated and characterized from *Thermoelectric Handbook: Macro to Nano*, whereas $ZT = 5$ suggests the future desire for TE materials. At $ZT = 5$ values, standalone thermoelectric could provide a similar performance to a practical vapor compression cycle in refrigeration. In the combined system, TE provides a minor improvement in efficiency, but plays a relatively significant role at lower interface temperatures. Unfortunately, based on our analysis, the theoretical combined system of TE ($ZT =1$) on top of a Rankine cycle could not achieve the best ever marked efficiency ($\sim$ 50%) for Brayton cycle + Rankine cycle. The engineering of TE materials with $ZT = 4 - 6$ is necessary to reach the best efficiency. Similar to the single cycle efficiency of Coal Rankine cycle, Solar Brayton cycle efficiency is also found to be $\sim$ 50%. Nevertheless, there is still room to put TE on top of them to enhance the efficiency.

6.1.9 *Energy economy analysis*

The energy production cost [\$/kWh] as the function of operation hours and interface temperature for the parts of TEG and ST is

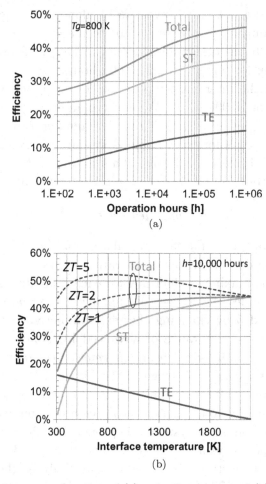

Fig. 6.7: Efficiency as a function of (a) operation hours and (b) interface temperature. (a) is a plot for an interface temperature of 800 K and (b) is a plot for 10,000 hours of operation with $ZT \sim 1$. The dashed curves show the cases of $ZT = 2$ and $ZT = 5$.

shown in Fig. 6.8 with variations of ZT values of 1, 2, and 5. There is a trade-off behavior between TE and ST for the interface temperature as expected. For the TEG, impact of enhancing the Z factor ($Z = \sigma S^2/\beta$) is separated by the change in power factor (σS^2) indicated by dotted curves and the change in thermal conductivity (β) indicated by dashed curves. Energy cost of $ZT = 5$ approaches ST.

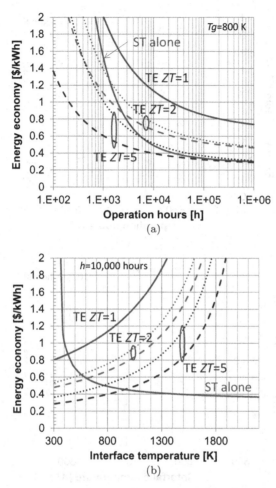

Fig. 6.8: Energy cost of ST only and TE only vs (a) operation hours and (b) interface temperature for 10,000 hours of operation with variations of the material properties of TE ($ZT = 1$, 2, and 5). Dotted curves show the impact of improving power factor and dashed curves show the impact of thermal conductivity.

Figure 6.9 shows the total energy production cost of the combined system. Longer operation always lowers the energy cost. The cost reduction contribution of the TE topping cycle is approximately 20% of the combined system at infinitely long operation hours. Impact of interface temperature is small and there is a very flat valley for the energy cost between 800 K and 1200 K. This trend shifts to

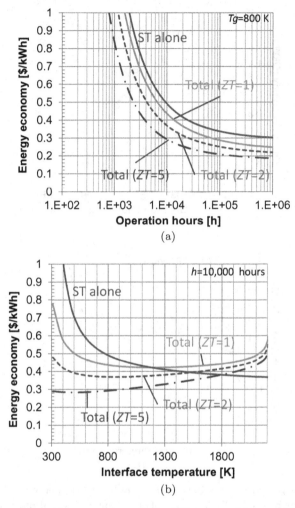

Fig. 6.9: Energy cost in comparison to the ST only vs (a) operation hours and (b) interface temperature with variations of ZT of TE ($ZT = 1$, 2, and 5).

a lower cost at a lower interface temperature with more advanced TE material. If compared to the most cost-effective technology ranging 0.1 $/kWh, the investigated system cannot compete as the first option may need additional consideration, e.g. further co-generation of low-grade heat recovery.

6.2   Waste Heat Recovery in Industrial Processes

There is a large potential for waste heat recovery in industrial manufacturing processes. Heat is one of the main critical energy inputs for
reforming materials. The thermal insulation component has a technological challenge because of the immense heat energy that needs to
be handled. In fact, the rejected or waste heat energy from industrial
process was 12.9 Quads (U.S., 2017) and about equal to the energy in
service [28]. This waste heat is around 13% of total energy consumption in the United States. The temperature of the waste heat is higher
than that from the transportation or commercial/residential building
sectors. Hence, the heat recovery from industrial process is crucial
from the viewpoint of both the thermodynamic quality and the total
energy volume. Steel, various metals, and glass reforming processes
are typical cases. A discussion in the following section focuses on
heat recovery from refractory walls as a component of a glass melting process. This is only a part of the entire glass reforming process
and requires the highest temperature of the material. The rest of the
potential heat recovery entrusts the analysis to the readers.

6.2.1   *Glass melt process*

Melting glass pellets requires a furnace with a temperature of over
1500°C for downstream glass shaping processes and hence a large
amount of exergy is available but currently destroyed. Due to the
high-temperature gradients, large amounts of parasitic heat loss exist
within the process.

A cross-sectional sketch of a glass melt process is shown in
Fig. 6.10. Four locations are initially identified for the location
of potential waste heat energy harvesting with thermoelectrics:
(a) crown ceiling of the glass melt pool, (b) sidewalls of the glass
melt pool, (c) fireports, and (d) the melt glass cooling in a pre-
forming process. Table 6.1 summarizes the available waste heat at
different locations. These heat fluxes are the maximum allowable
heat flux q_{limit} for the design of waste heat recovery while maintaining the fuel burning rate constant. The bottom wall of the pool is

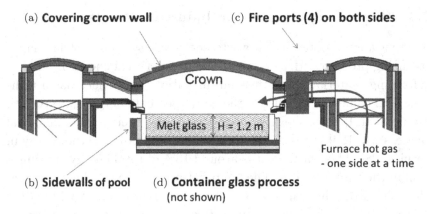

(a) **Covering crown wall**   (c) **Fire ports (4) on both sides**

(b) **Sidewalls of pool**   (d) **Container glass process**
(not shown)

Fig. 6.10:   Cross-sectional sketch of glass meting pool, indicating locations of potential waste heat recovery.

Table 6.1:   Heat loss estimated for four potential locations.

Type	Location	Area [m²]	Heat flux [kW/m²]	Loss [kW]	Ratio*	Cooling method available
(a)	Crown	200	3.5	700	5.6%	Air
(b)	Sidewall	130	2.8	364	2.9%	Air
(c)	Fireport	38.4	9.0	346	2.8%	Air $\rightarrow$ Water
(d)	Cooling	40	5.9	236	1.9%	Air

Note: *Ratio is the heat loss relative to the tangible heat demand for melting glass pellets.

excluded since the integration could be overly challenging. Melted glass from the pellets must maintain a temperature of 1500°C in the pool, while the furnace hot gas temperature exceeds that temperature [29]. Although the pool is built with thick refractory walls [30], heat loss per unit area is in the range of 9 kW/m² at fireports and around 3 kW/m² at other walls [31]. The heat demand is 12.5 MW for melting glass pellets in 500 ton/day (5.8 kg/s) production based on 2172 kJ/kg of tangible heat for melting pellets at 1500°C [32].

Among the variations of thermal paths from the melted glasses, the fireports exhibit the largest heat density, which yields the best potential for lowest cost heat recovery with thermoelectrics by partially implementing it within the refractory wall, while the total heat removal for the temperature control is maintained. The analytic investigation of this waste heat recovery follows.

A fireport is a channel allowing the hot gas to escape through it. Refractory walls of the fireport are designed to maintain the hot gas at 1500°C. The following sketches show the potential integration of a TEG at the locations of interest. High-temperature gradients across the thermoelectric generator requires a water-cooling heat sink to process the large amount of heat. The investigation is based on a typical thermoelectric figure of merit $(ZT = 1)$, which is designed optimally for a 500 ton/day (5.8 kg/s) scale glass production.

6.2.2   *Optimization for high-temperature waste heat recovery*

The refractory wall is typically made of aluminum−zirconia−silica (AZS) alloys [33]. The AZS wall is partially replaced with the TEG modules whose thickness varies from 0 m up to 0.54 m. Water-cooled heat sink is integrated at the outside wall, as shown in Fig. 6.11(c). TEG is designed to match the reduced thermal resistance by thinning the AZS wall in order to keep the heat flow constant. The TEG can be integrated at either the cold side or hot side of the AZS wall, as shown in Fig. 6.12. An average thermal conductivity of 3.5 W/(m K) is used for the AZS based on the temperature-dependent data [34].

Between the hot gas and the refractory wall, an effective heat transfer coefficient of 200 W/(m^2 K) is considered as previously described. The thermal resistance of the current refractory wall is estimated to be 9.7 K/W for each fireport. An effective heat transfer coefficient for water cooling at 40°C is determined as 1500 W/m^{2}K with a tube diameter of 0.01 m with a pitch of 0.015 m in a copper cold plate covering the perimeter of the fireport. The heat transfer coefficient is calculated based on mass flowrate of 0.021 kg/s per tube with thermofluid-dynamic correlations between the Reynolds number and the Nusselt number.

Thermoelectric material property values are 1.5 W/(m K) of thermal conductivity and 56,000 1/(Ω m) of electrical conductivity, while the Seebeck coefficient is adjusted to match the average figure of

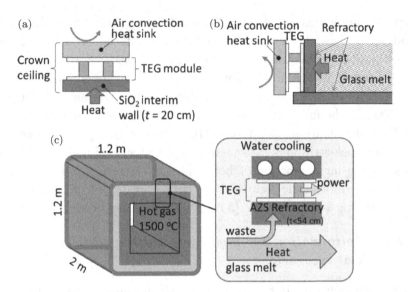

Fig. 6.11:   Sketch of implementations of TEGs in a glass melt process: (a) crown ceiling wall with air convection heat sink, (b) side refractory wall of glass melt pool with air convection heat sink and (c) one of the four fireports with water-cooling heat sink (with circular channels).

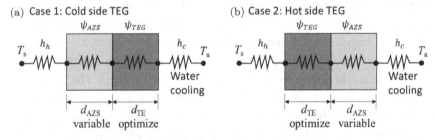

Fig. 6.12:   Thermal model for optimizing the thermoelectric leg thickness integrated with refractory AZS wall. The heat flux is maintained at 9 kW/m^2.

merit ZT to be unity independent of the temperature and the design condition.

Substrates, packaging, and electrical contact are included in the TEG model and cost calculations. Fill factor of the thermoelectric module is selected to be 10%. The module substrates are made of alumina (Al$_2$O$_3$) with a thickness of 1×10^{-3} m for both the hot and cold sides. A hermetic packaging with moderate vacuum gap is assumed for the module. The vacuum gap reduces the parasitic heat loss as well as prevents from the long-term performance degradation. The effective thermal conductivity of the air gap is 4×10^{-3} W/(m K) based on a moderate vacuum packaging at 1 kPa of air pressure and the specific contact resistivity is assumed to be 10^{-6} $\Omega\,\mathrm{cm}^2$. Two thermal model cases are investigated for the integration of thermoelectric with refractory AZS wall. Case 1 considers the AZS wall for the hot gas side and the thermoelectric is on the cold side. Case 2 reverses the order. The temperature range for the thermoelectric heat recovery is very different between these two cases in order to see the impact for energy conversion. In this investigation, using exergy [35] as the measure makes the visibility clearer.

In order for the heat flux to be limited to 9 kW/m^2, power output is not maximized and instead the thermal resistance is adjusted by the thickness of thermoelements for these cases. There are limitations of design, where the actual heat flux cannot go beyond the heat flux constraint with given conditions for thicker AZS design. The exergy of both cases begins at the same point when AZS wall thickness is zero. The total specific exergy rate including the hot side irreversible heat loss is found as $Ex_{in} = (1 - T_a/T_s)q$. Following is the expression of specific exergy rate through the thermoelectric system Ex_h J/(m^2 s):

$$Ex_h = \left(1 - \frac{T_a}{T_h}\right) q \qquad (6.21)$$

Similarly, the following equation shows the specific exergy rate available for thermoelectric energy conversion EpJ/(m^2 s), respectively, at the temperature T_h:

$$Ep_{TE} = \left(1 - \frac{T_c}{T_h}\right) q \qquad (6.22)$$

On the other hand, the power output w_{TE} W/m^2 by the thermoelectric energy conversion is

$$w_{TE} = \eta q \qquad (6.23)$$

where

$$\eta = \left(1 - \frac{T_c}{T_h}\right)\left(\frac{\sqrt{1 + ZT} - 1}{\sqrt{1 + ZT} + T_c/T_h}\right) \qquad (6.24)$$

as T_h and T_c are determined and q is an available heat flux applied to the TEG. The Ep_{TE} exergy rate available for energy conversion for Case 1 is very similar to the Ex_h exergy rate through the system since the cold side temperature T_c for Case 1 is very close to the water (cold reservoir) temperature T_a. The harvested energy is clearly larger in Case 1, which is a favorable solution for this particular study as the heat loss from the wall still remains constant. Figure 6.13 shows the performance as a function of the AZS refractory wall thickness.

A challenge for the implementation of high-temperature thermoelectric module is the limited technology options in making the TE material and metal contacts that survive very high temperatures at the hot side of the TEG elements. The temperature is lower than the hot gas (for glass melting), but still it could limit the design of TEG, requiring it to keep the part of AZS refractory wall. Assuming that

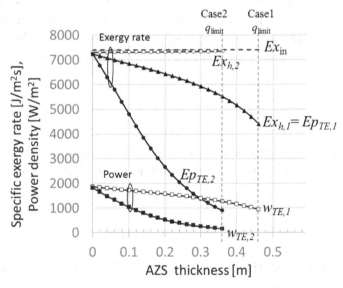

Fig. 6.13: Specific exergy rate Ex [J/(m² s)], thermoelectric energy conversion potential Ep [W/m²], and power density w_{TE} [W/m²] vs AZS thickness [m] for Case 1: cold side TEG and Case 2: hot side TEG.

a half of AZS wall thickness is replaced by TEG, power output is 13.9 kW for each fireport from the available surface area of 9.6 m^2, this limits the highest temperature T_h of TEG elements to ~740°C according to the calculation based on the above model. With a total of four fireports, 55.6 kW of electricity could be generated from the waste heat from a 500 ton/day (5.8 kg/s) glass processing facility.

Since this implementation maintains the same outgoing heat flux as the current fireport air-cooling solution, extra fuel is not needed while electricity is generated with efficiencies exceeding 15%. In addition to functional electrothermal optimization, high-temperature gradient thermomechanical optimization for maximum robustness, stable high-temperature contacts and thermoelectric material stability are equally important for high-temperature waste heat recovery.

6.3   Waste Heat Recovery from Vehicle Exhaust Gas

Heat rejected or waste heat energy from transportation segment was 22.2 out of 28.1 Quads (U.S., 2017). Not only the energy utilization factor is as low as 21%, approximately 1/3 of unused primary energy comes from this segment. The waste heat comes typically from internal combustion Diesel, Otto, or Brayton cycle heat engines. The major heat flow includes exhaust hot gas, cooling heat rejection through radiators, convection heat loss from the engine body, and also some other paths contribute. Hence, exhaust gas heat recovery is not the only solution for this big challenge. On the other hand, it is the most preferable heat path to apply for heat energy recovery technologies. Engine exhaust gas contains relatively high temperature and the flow is bundled for filtering toxic gasses. In this section, gas turbine (Brayton cycle) is excluded since the exhaust of the engine is mostly used as source of power for propulsion in airplanes or used as the steam generator for the bottoming Rankine cycle in power plants.

6.3.1   *Vehicle exhaust heat recovery*

The uniqueness of automotive exhaust from the internal combustion engine is the requirement for the precise design of the pressure

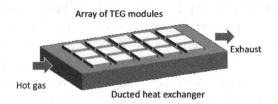

Fig. 6.14: Exhaust gas heat recovery system configuration investigated by Kumar *et al.* [38]. Thermoelectric generator (TEG) modules are located on both sides of the rectangular channel heat exchanger for flue gas to go through.

loss in the exhaust system, which is known as 'backpressure'. If the pressure load through the exhaust increases, the engine shaft power will be reduced. The main objective of utilizing a thermo-electric power generator in an exhaust gas heat recovery system (Fig. 6.14) is to save fuel consumption throughout the entire power system because the harvested electricity will be used anyway. Recent heavily electrified vehicle requires a significant amount of electricity, which has been supplied only by the alternator in typical non-electric vehicles. The alternator consists of an electric magnetic motor mechanically connected to the engine power shaft with a revolution speed adjustment. There have been several practical studies reported with prototyping the waste heat harvesting, e.g. U.S. Department of Energy promoted developments among private companies [36]. Fossil fuels are widely used to operate the vehicles because of their high energy density of nearly 40 MJ/l. In typical driving conditions, useful mechanical thrust power is extracted in primary energy efficiency of 18−30%. The primary energy efficiency stands for the efficiency of conversion to useful energy from the chemical energy contained in the fuel. In contrast, the theoretical thermodynamic energy efficiency of the engines ranges 30−40%. Hence, there exists an additional loss. Nearly 40% of energy dissipates via radiator (air-cooling exchanger for engine coolant) and the major part (30%) of the remaining heat is exhausted via flue gas. A prototype of the waste heat recovery device in the range of 700 W of power output has been developed and tested in several variations of geometrical configurations, such as Longitudinal, Transverse, Hexagonal, and Cylindrical [37, 38].

Major challenge of waste heat recovery from the exhaust gas is the large fluctuation of flow rate and temperature of the gas depending on the engine operations. The variation makes the design optimization difficult and results in the lower performance in fuel saving than the expected level based on a typical engine operation.

6.4 Solar Energy Harvesting

This section highlights the harvesting of solar energy. To begin with, the theoretical upper limit of photothermal conversion is briefly discussed. Since solar energy conversion is a thermodynamic process, it obeys the second law of thermodynamics. Then, heat energy conversion is discussed followed by the combined heat and power (CHP) generation in the last part.

6.4.1 *Solar energy*

Solar irradiation is considered as the energy transfer from an infinitely large heat source with a spectrum that has the peak temperature of 5762 K according to the Planck law. From the distance from the sun and the limited diameter of the sphere, mean planet temperature of the earth T_p is calculated as 288 K. In the photothermal energy conversion, theoretical maximum conversion efficiency η of solar harvesting device is limited up to the Carnot efficiency $\eta_{\text{Carnot}} = (1 - T_p/T_s)$ by following the second law of thermodynamics. Optical concentration of solar irradiation on terrestrial determines the receiver temperature. Using optical concentration C, the net incoming energy per unit surface area q is found as

$$q = Cf\sigma T_s^4 + (1 - Cf)\,\sigma T_p^4 - \sigma T_b^4 \tag{6.25}$$

where $f = r^2/R^2$ with r is radius of earth and R is radius of sun. T_b is the temperature of surrounding earth and hence a part of the energy is emitted to the sky. It is natural to take an approach to block the emission for a part of the spectrum at energy E_g to prevent the loss. This energy level E_g is related to the threshold of photon−electron interaction in semiconductor, called bang gap in photovoltaic materials. Considering the spectrum of the energy using Bose−Einstein distribution function, the net incoming energy flux becomes

$$q = Cf \int_E g^\infty \frac{E^3 dE}{\exp(E/k_B T_s) - 1} + (1 - Cf) \int_E g^\infty \frac{E^3 dE}{\exp(E/k_B T_p) - 1}$$

$$- \int_E g^\infty \frac{E^3 dE}{\exp(E(1 - \eta)/k_B T_p) - 1} \tag{6.26}$$

where η is the energy conversion efficiency to satisfy the relation. At the limit $C \rightarrow 46300, T_p \rightarrow T_s$ and then it maximizes the efficiency.

6.4.2   *Solar thermo-photovoltaic system*

This approach uses a flat absorber to absorb sunlight on one side and then emit the radiation from the other side to the flat surface of a photovoltaic device spaced with a specific gap. The absorber converts the spectrum into a longer wavelength due to the lower temperature compared to the original solar spectrum. A selective absorber reduces the emission back to the sky (due to the high concentration) and improves the performance. Investigation into the materials and layer structure has been conducted [39].

As an approach to reduce the transmission loss to the converter, near-field radiation heat transport is actively studied [40,41]. The transmittance of energy significantly increases as the distance decreases between emitter and energy converter. In principle, the transmission loss could approach zero by minimizing the gap. However, the retention to perfectly align the flat surfaces with nanometer gap spacing is a significant challenge due to the van der Waals forces as well as the required fabrication precision.

6.4.3   *Solar thermal tower system*

Large-scale concentrated solar thermal tower systems have been developed in the last few decades mostly in desert or unpopulated regions, according to Kolb [42] and Caldés [43]. The typical size of the footprint from these reflectors placed on the ground is 1,000–10,000 m^2 and the incoming peak energy is 10–100 MW. The array of mirror reflectors is synchronized to track the light from the sun so as to concentrate it to the point at the top of the tower with a concentration of about 1,000 suns. This creates temperatures high enough for molten salt or even directly generates the superheated steam. The energy conversion is performed by a superheat Rankine cycle with a steam temperature of about 850 K and the efficiency of electric power generation is about 40%. Due to the high pressure, over 25 MPa, a higher temperature significantly increases the cost for an improvement in efficiency of only a few percent. Hence, the thermodynamic limitation is based on this temperature which is, in practice, difficult to exceed.

6.4.4 *Solar trough system*

The simplest way to use solar energy is to convert it to heat. The heat is used for space heating or hot water supply and sometimes kept for later use with thermal storage. Simply, the system uses a reflective mirror to concentrate the sunlight onto a water pipe to absorb the energy as heat. The energy conversion efficiency depends on the level of concentration. A dish concentrator provides higher concentration and thus provides higher energy efficiency than that of a parabolic trough, but the cost of a parabolic trough is much less than the dish due to the geometrical complexity differences. The efficiency will be closer to the upper bond thermodynamic efficiency rather than other systems generating electricity, but the exergy is completely lost. Considering the fact that real life uses electricity significantly for heating or cooling, heat energy generation is still essential. Solar heating systems are helpful in regions with no electricity supply. Even for developed countries, solar cooling systems are important for energy sustainability.

6.4.5 *Solar thermoelectric*

Currently, commercialized thermoelectric (TE) solar energy conversion is limited. A unique device is demonstrated with practical conversion efficiency at 4.6% [44] under a 1 kW/m^2 solar irradiation. Through analytical modeling and electrothermal co-optimization [45], the theoretical efficiency at the maximum power output of a system with simple symmetric contacts is found with figure of merit Z value as

$$\eta_{\mathrm{TE}} = \eta_{\mathrm{abs}} \frac{W_{\mathrm{max}}}{Q} = \eta_{\mathrm{abs}} \frac{1}{Q} \left(\frac{Z}{4 \left(m + 1 \right)^2 \sum \psi} \left(T_s - T_p \right)^2 \right) \quad (6.27)$$

where m is the ratio of the TE internal electrical resistance over the load resistance, $\Sigma \psi$ is the sum of external thermal resistance in the system, and η_{abs} is the absorber efficiency, which can be as high as 84.5% if a perfect selective absorber is readily available. The selective absorber stands for the absorber with the surface fabricated to maximize the absorption in an energy band of infrared wavelength (typically shorter than 3 μm) and maximize the reflection (poor emission) in far-infrared wavelength (typically in range of 10 μm). If the material is ideally adaptable for such high temperatures, a material with $Z = 3 \times 10^{-3}$ (1/K) as an ordinal value for room temperature could yield an efficiency of 40.78% with symmetric contacts. It varies from 34.5% for zero contact resistance to the cold side to 41.2% for zero contact resistance to the hot side. The overall system efficiency is found to be 84.5% × 34.8% = 29.4%.

6.4.6   *Combined heat and power generation*

Solar irradiation is renewable, clean, and cost-free resource and has been typically expended for heating up water. As discussed in the earlier sections, some systems generate steam or vapor to drive Rankine cycle and others boil the water for space heating.

In both ways of harvesting the solar energy, there still remains a big exergy gap between the source temperature and the temperature actually used. Similar to topping power generation, thermoelectric can be used as the high-temperature cycle on top of the current technology. Looking at the system from the other side, thermoelectric generator wastes the majority of heat energy due to its relatively moderate efficiency. The lower temperature side of TEG must reject heat anyways. The combined heat and power generation provides high energy efficiency in systems utilizing solar harvesting.

In order to maintain the required temperature to build the system, optical concentration of solar is crucial. The natural solar irradiation on the ground at sea level is only about 1 kW/m^2, while thermoelectric is cost effective with a heat source of upper 100 kW/m^2. Hence, a concentration of about 100 times should work well for an economical power generation.

The following shows an experimental investigation of combined heat and power, demonstrating with some off-the-shelf components, such as a Fresnel lens for optical concentration and an aqua pump for circulating water.

The electrical load resistor is always optimized for the maximum power output from the TEG, while the thermal resistance is supposed to be the design variable. Due to limited availability of thermoelectric modules off the shelf, however, thermal impedance of the module is not optimized. Therefore, concentration and flow rate are the variables.

Also, inhomogeneous power generation for each thermoelement in a module resulted in significant degradation to power output. The inhomogeneity can be improved by appropriate heat spreader on the hot side. It would be further aligned if optical path design modification is allowed for the Fresnel lens.

As a result, harvested combined heat and power observed more than 53% of the received solar energy with the thermally non-optimized module, while 0.44–0.46 W of electricity was generated by the TE module with a center temperature difference of 143–152 K between the hot side and the cold side in the module. If the TEG module would be thermally optimized, 10 times of electricity could be generated.

6.4.7   *Solar energy corrector — a design case*

The diagram of the experimental apparatus is shown in Fig. 6.15. TEG is a commercially available module consisting of bismuth telluride alloy materials between two ceramic (aluminum oxide) plates. The module has dimensions of 30 mm × 30 mm with a thickness of 3.4 mm and 254 elements connected electrically in series and thermally in parallel. Both modules contain elements

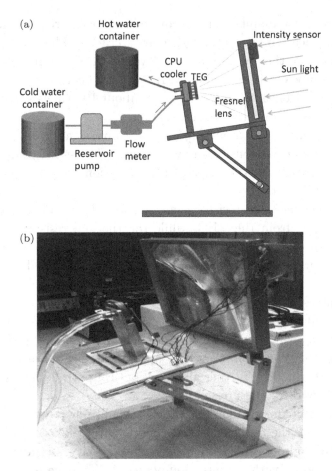

Fig. 6.15: (a) Diagram of the system. (b) Final construction with 86x Fresnel lens. The overall system stands approximately 600 mm tall while its base covers 300 mm × 300 mm area. Fixing the TE and CPU cooler together is an adjustable stage, which can reposition to match the focus of each of the two tested Fresnel lenses.

of dimensions 1 mm × 1 mm for the cross-section with a length of 1.83 mm.

6.4.7.1   *Fresnel lens*

A Fresnel lens fits smaller scale solar concentration instead of parabolic solar dish collectors [46] and is readily available in the market. Especially, relatively low-magnification plastic lenses were quite well developed for overhead projectors (OHP) since the 1960s.

One of the off-the-shelf Fresnel lenses is utilized for sunlight concentration. The component is composed of optical grade acrylic with 92% transmission of wavelengths ranging from 400 to 1100 nm. Near-infrared (IR) wavelength is also transparent. This lens is transparent for approximately 85% of the solar spectrum as the overall range is 300−2500 nm, so that incoming energy is reduced. An 86x concentration Fresnel lens with dimensions of 279 mm × 279 mm and 2.2 mm thick with a focal length of 178 mm is used. The hot side of the TEG exposed to solar radiation is painted black to give a larger emissivity at around 0.85 and maximize the absorption of incoming heat energy.

6.4.7.2 *Solar tracker*

The system employs a solar intensity sensor. The orientation angle to the sun was manually adjusted every 5 min by hand. With additional mechanical figure and motor drives, horizontal and vertical angle tracking with circuit is possible as a future extension.

6.4.7.3 *Water pump*

An electric powered pump compatible with the CPU cooler is applied to the cold side of TEG. It supplies the water at the flow rate of 500 ml/min. Absorbed heat Q_c is calculated from the temperature difference between the inlet and outlet with properties at room temperature:

$$Q_c = \rho \dot{V} C_p \Delta T \tag{6.28}$$

6.4.7.4 *Cooler and the heat transfer*

On the back side of the TEG, a commercial water flowing CPU cooler is used to extract heat from the cold side. This copper-based CPU cooler contains 100 rectangular pillars with a 1.5 mm × 1.5 mm cross-section and a 4 mm height (Fig. 6.16). The thermal resistance of the CPU cooler based on the given flow rate is calculated by knowing

Fig. 6.16: Cross-section of heat pin-fin heat exchanger. 1.5 mm × 1.5 mm square cross-section and 4 mm height are common for each of the 100 pins.

Fig. 6.17: Diagram of the coolant flow. Section 6.1 indicates the impinging flow and Section 6.2 indicates the parallel flow.

the effective heat transfer coefficient (Fig. 6.17). According to the fin design, the effective heat transfer coefficient is determined for 0.75 cm^2 of pin-fin area following Kondo and Matsushima [47].

Overall effective heat transfer coefficient is determined by Eq. (6.29). This is translated to 0.2 K/W of the thermal resistance:

$$h_f = \frac{A_1 h_1 Fr_L + A_2 h_2 Fr_C}{A_f} \qquad (6.29)$$

where A_1 and A_2 are the surface area of fins, h_1 and h_2 are the heat transfer coefficients, and Fr_L and Fr_C are the geometry-related factors. A_f is the bottom surface area and h_f is the effective heat transfer coefficient at the area A_f.

6.4.7.5   Electrical load

Before running the solar harvesting test, the electrical load resistance is optimized according to $R_L = \sqrt{1 + ZT}R$, as ZT value is 0.943. Estimated optimum load resistances for the TEG in test is 7.91 Ω.

6.4.7.6   Experimental results

The experiments are conducted in 5-min intervals. This time duration is sufficient to have the temperatures stable. This dataset was acquired under clear sunshine weather at Santa Cruz, California (latitude = North 36.98°) in early July 2011. The air temperature was 22–23°C during the experiments.

Using the 86x Fresnel lens to concentrate solar radiation onto the TE, the CPU cooler maintained the cold side of the TE at 45°C. With a 142.5°C temperature difference, the TEG generated 0.44 W across an 8.1 Ω load resistor. Assuming that the solar radiation has a power flux of 650 W/m² based on the chart [48], the effective area of the studied Fresnel lens 0.066 m² receives 42.7 W of solar power to convert. In reality, experiments show that the system generates 0.44 W of electrical power and recovers 21.5 W of heat. Therefore, the system efficiency is found to be 51.4% (Fig. 6.18). The efficiency of electric power alone is 1%, while the predicted efficiency is 5.9% based on the experimental temperatures T_h and T_c and calculated following the well-known equation, Eq. (6.30), following Ref. [49],

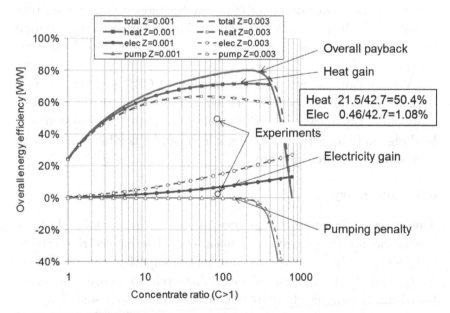

Fig. 6.18: Efficiency as a function of optical concentration from the model based on the properties on experiments and measured data points. Efficiency includes heat and power co-generation with variations of effective thermoelectric Z values of 0.001 (minimum off the shelf) and 0.003 (advanced material).

where the T_h is the hot side temperature and T_c is the cold side temperature of the TE module:

$$\eta_{TE} = \frac{(m-1)\,(T_h - T_c)}{(mT_h + T_c)} \tag{6.30}$$

Thermal impedance is unoptimized in this experiment. In addition, a major cause of this discrepancy is mostly induced by nonuniformity of the optical input on to the rectangular surface of the thermoelectric generator. This causes the actual temperature difference across each thermoelement.

6.4.8 *CHP system optimization*

From 42.7 W of incoming solar radiation on approximately 66 cm^2 area, 0.46 W of electricity was produced by the TE module. The waste heat also recovered 21.5 W. In a realistic application, water flow rate is an important factor to obtain the most useful temperature of hot water. In the model-based calculation with a custom design of a thermoelectric module, harvesting 10 times electrical power is possible.

A significant benefit of a thermoelectric CHP generator comes from its scalability. Since Fresnel lenses are inexpensive and widely available in the market with thermoelectric modules, the system can be designed in an array structure, which can be flexible for size with very similar cost per unit area.

6.5 Summary

This chapter discussed the analysis of specific industrial-scale thermoelectric power generation applications. Thermoelectrics can be considered to be attached at the heat dissipation site of a thermodynamic heat engine as a bottoming cycle to recover the waste heat from the engine. Exhaust gas heat recovery and solar harvesting co-generator with photovoltaic system are described as other important examples. Contrary to this conventional wisdom, however, the topping cycle of the heat engine can also be an ideal utilization due to the large amount of exergy wasted when the heat from a fuel-burning boiler is transferred to the steam. Due to the moderate efficiency of thermoelectric generators, the impact of applying a retrofitting to an existing power generator in relatively lower temperature range can be small, but still adds a useful range of power output, e.g. additional 5% to the coal-fired Rankine cycle power generator with an efficiency of 40%. Another big benefit of the topping cycle is that it can be significantly more cost effective when compared to the bottom-cycle waste heat recovery system. This happens due to the thinner optimum leg for high heat flux as discussed in the earlier sections with analytic modeling and optimization.

References

[1] *New York Times* on-line topic, *Nuclear Energy*, Available at: http://topics. nytimes.com/top/news/business/energy-environment/atomic-energy/index. html, October 12 (2012).

[2] Lawrence livermore national laboratory, Energy flow chart — Estimated Energy Use in 2011: 97.3 Quads, Available: https://flowcharts.llnl.gov/ (2012).

[3] C. A. Parsons, *The Steam Turbine*, Cambridge University Press (1911).

[4] W. H. Wiser, *Energy Resources*, Springer (2000).

[5] F. Curzon and B. Ahlborn, Efficiency of a carnot engine at maximum power output, *Amer. J. Phys.*, **43**(1), 22 (1975).

[6] C. B. Vining, An inconvenient truth about thermoelectrics, *Nat. Mater.*, **8**, 83–85 (2009).

[7] S. Imano, E. Saito, J. Iwasaki and M. Kitamura, High-temperature steam turbine power plant, US Patent Application 20080250790.

[8] M. Chen, H. Lund, L. A. Rosendahl and T. J. Condra, Energy efficiency analysis and impact evaluation of the application of thermoelectric power cycle to today's CHP systems, *App. Ene.*, **87**(4), ISSN 0306-2619, DOI: 10.1016/j.apenergy.2009.06.009, 1231–1238 (2010).

[9] L. Chen, J. Li, F. Sun and C. Wu, Performance optimization of a two-stage semiconductor thermoelectric-generator, *Appl. Energy*, **82**(4), ISSN 0306-2619, DOI: 10.1016/j.apenergy.2004.12.003, 300–312 (2005).

[10] T. Caillat, J.-P. Fleurial, G. J. Snyder and A. Borshchevsky, Development of high efficiency segmented thermoelectric unicouples, *Proceedings of ICT2001*, 282–285 (2001).

[11] K. Qiu and A.C.S. Hayden, Integrated thermoelectric and organic Rankine cycles for micro-CHP systems, *Appl. Energy*, **97**, 667–672 (2012).

[12] X. Gou, H. Xiao and S. Yang, Modeling, experimental study and optimization on low-temperature waste heat thermoelectric generator system, *Appl. Energy*, **87**(10), 3131–3136 (2010).

[13] J. W. Fairbanks, Vehicular thermoelectrics: A new green technology, *Proceedings of the 2nd Thermoelectrics Applications Workshop* (2011).

[14] Q. E. Hussain, D. R. Brigham and C. W. Maranville, Thermoelectric exhaust heat recovery for hybrid vehicles, *SAE Int. J. Eng.*, **2**(1), 1132–1142 (2009).

[15] J. W. LaGrandeural, L. E. Bellal and D. T. Crane, Recent progress in thermoelectric power generation systems for commercial applications, *MRS Proc.*, **1325** (2011).

[16] D. Kraemer, B. Poudel, H.-P. Feng, J. C. Caylor, B. Yu, X. Yan, Y. Ma, X. Wang, D. Wang, A. Muto, K. McEnaney, M. Chiesa, Z. and G. Chen, High-performance flat-panel solar thermoelectric generators with high thermal concentration, *Nat. Mater.*, **10**, 532–538 (2011).

[17] K. Yazawa and A. Shakouri, System optimization of hot water concentrated solar thermoelectric generation, *Proceedings of the 3rd International Conference on Thermal Issues in Emerging Technologies Theory and Applications (ThETA3)*, 283–290 (2010).

[18] J. Xiao, T. Yang, P. Li, P. Zhai and Q. Zhang, Thermal design and management for performance optimization of solar thermoelectric generator, *Appl. Energy*, **93**, 33–38 (2012).

[19] K. Yazawa and A. Shakouri, Optimization of power and efficiency of thermoelectric devices with asymmetric thermal contacts, *J. Appl. Phy.*, **111**(2), 024509 (2012).

[20] K. Yazawa and A. Shakouri, Optimizing cost-efficiency trade-offs in the design of thermoelectric power generators, *Envi. Sci. and Tech.*, **45**(17) 7548–7553 (2011).

[21] K. Yazawa and A. Shakouri, Cost-effective waste heat recovery using thermoelectric systems, *Invited Feature Paper, J. Mat. Res.*, **27**, 1–6 (2012).

[22] C. B. Knowles and H. Lee, Optimized working conditions for a thermoelectric generator as a topping cycle for gas turbine, *J. of Appl. Phy.*, **112** (2012).

[23] D. M. Rowe (Ed.), *Thermoelectric Handbook Macro to Nano*, CRC Press, 83–142 (2006).

[24] A. Shakouri, Recent developments in semiconductor thermoelectric physics and materials, *Annu. Rev. Mater. Res.*, **41**, 399–431 (2011).

[25] D. L. Nika, E. P. Pokatilov, A. A. Balandin, V. M. Fomin, A. Rastelli and O. G. Schmidt, Reduction of lattice thermal conductivity in one-dimensional quantum-dot superlattices due to phonon filtering, *Phys. Rev., B*, **84**, 165415 (2011).

[26] H. Bauer (Ed.), *Automotive Handbook 4th Edition*, Robert Bosch GmbH, 238–239 (1996).

[27] A. Van Maaren, D. S. Thung and L. R. H. De Goey, Measurement of flame temperature and adiabatic burning velocity of methane/air mixtures, *Comb. Sci. Tech.*, **96**(4–6), 327–344 (1994).

[28] The US Energy Flow Chart 2017, Lawrence livermore national laboratory, Available: https://flowcharts.llnl.gov/commodities/energy.

[29] S. Aoki, S. Kawachi and M. Sugeno, Application of fuzzy control logic for dead-time processes in a glass melting furnace, *Fuzzy Set. Syst.*, **38**(3), 251–265 (1990).

[30] M. Hubert, Lecture 3: Basics of industrial glass melting furnace, IMI-NFG *Course on Processing in Glass* (2015).

[31] L. Pilon, G. Zhao and R. Viskanta, Three-dimensional flow and thermal structures in glass melting furnaces. Part I. Effects of the heat flux distribution, *Glass Sci. Technolog.*, **75**(2), 55–6 (2006).

[32] M. Hubert, Lecture 2: Industrial glass melting and fining processes, IMI-NFG *Course on Processing in Glass* (2015).

[33] T.-J. Wang, Modelling of fused cast alumina refractory, *Br. Ceram. Trans.*, **98**(2), 62–70 (1999).

[34] S. L. Cockcroft, J. K. Brimacornbe, D. G. Walrod and T. A. Myles, Thermal stress analysis of fused-cast AZS refractories during production: Part II, Development of thermo-elastic stress model, *J. Amer. Cera. Soc.*, **77**(6), 1512–1521 (1994).

[35] G. Wall, *Exergy: A Useful Concept*, Chalmers Tekniska Hogskola (1986).

[36] Heat recovery from automotive vehicles (DOE), Available: https://www. energy.gov/eere/vehicles/downloads/thermoelectric-waste-heat-recovery-pro gram-passenger-vehicles-0.

[37] S. Kumar, S. D. Heister, X. Xu, J. R. Salvador and G. P. Meisner, Thermoelectric generators for automotive waste heat recovery systems part I: Numerical modeling and baseline model analysis, *J. Elect. Mater.*, **42**(4), 665–674 (2013).

[38] S. Kumar, S. D. Heister, X. Xu, X., J. R. Salvador and G. P. Meisner, Thermoelectric generators for automotive waste heat recovery systems part II: parametric evaluation and topological studies, *J. Electron. Mater.*, **42**(6), 944–955 (2013).

[39] P. Bermel, K. Yazawa, J. L. Gray, X. Xu and A. Shakouri, Hybrid Strategies and Technologies for Full Spectrum Solar Conversion, *Energy Environ. Sci.*, **9**(9), 2776–2788 (2016).

[40] L. Hu, A. Narayanaswamy, X. Chen and G. Chen, Near-field thermal radiation between two closely spaced glass plates exceeding Planck's blackbody radiation law, *Appl. Phy. Lett.*, 92, 133106 (2008).

[41] E. Rousseau, A. Siria, G. Jourdan, S. Volz, F. Comin, J. Chevrier and J.-J. Greffet, Radiative heat transfer at the nanoscale, *Nat. Phot.*, **3**, 514–517 (2009).

[42] G. J. Kolb, D. J. Alpert and C. W. Lopez, Insights from the operation of solar one and their implications for future central receiver plants, *Solar Energy*, **47**(1), 39–47 (1991).

[43] N. Caldés, M. Varela, M. Santamaría and R. Sáez, Economic impact of solar thermal electricity deployment in spain, *Ener. Pol.*, **37**(5), 1628–1636 (2009).

[44] D. Kraemer, B. Poudel, H.-P. Feng, J. C. Caylor, B. Yu, X. Yan, Y. Ma, X. Wang, D. Wang, A. Muto, K. McEnaney, M. Chiesa, Z. Ren and G. Chen, High-performance flat-panel solar thermoelectric generators with high thermal concentration, *Nat. Mat.*, **10**, 532–538 (2011).

[45] K. Yazawa and A. Shakouri, Optimization of power and efficiency of thermoelectric devices with asymmetric thermal contacts, *J. Appl. Phys.*, **111**, 024509 (2012).

[46] S. A. Kalogirou, Solar thermal collectors and applications, *Prog. Energy Combust. Sci.*, **30**, 231–295 (2004).

[47] Y. Kondo, H. Matsushima, Prediction algorithm of thermal resistance for impingement cooling of heat sinks for LSI packages with pin-fin arrays, *Heat Trans. Japan Res.*, **25**(7), 434–448 (1996).

[48] The renewable resource data center, *National Renewable Energy Laboratory, U.S. Solar Radiation Resource Maps* (1961–1990). Available: http://rredc. nrel.gov/solar/old_data/nsrdb/1961-1990/redbook/atlas/.

[49] D. M. Rowe, General principles and basic considerations, in *CRC Handbook of Thermoelectrics*, D. M. Rowe (Ed.), pp. 1–10 (2006).

Chapter 7

Wearables and Internet of Things

This chapter specifically discusses the thermoelectric energy conversion applications focusing on wearable devices and Internet of things (IoT). These applications share a lot of characteristics, such as room-temperature operation, small temperature difference, and low heat flux. Curved surfaces of the heat source are another similarity of both applications. Thermal contact is a big challenge in this case, and it is important to maintain the external thermal resistance as small as possible since the heat flux is very small. In addition to the flow bypass and thermal bypass concepts, flexible thermoelectric modules are discussed as an approach to obtain a good thermal contact between the TE module and the heat source. Fundamentals are discussed in the early part of the chapter, and then some unique characteristics of each application is specifically discussed in the later sections. As the Internet becomes available to connect remote sensors and actuators to the remote server, many unique features become possible with newly developed software and wireless communication. Thermoelectric heat energy recovery is a suitable technology for powering the necessary electronics by using on-site waste heat adjacent to the device. In wearable applications, two different types of thermoelectric energy conversion are considered: one is body heat harvesting and another is thermotherapy for both heating and cooling of the local skin.

7.1 Thermal Energy Harvesting for IoT

This section discusses a case study of a thermoelectric power generator for powering an acoustic sensor on water service pipelines and wirelessly transmitting signals for central monitoring. In order to harvest a temperature difference, there must be two different temperature fluid flows. It is popular to have two or more pipes placed in parallel in a commercial building, such as hot water for bathrooms and chilled water for computer rack cooling, etc. In order to maintain these pipes in healthy condition, monitoring the acoustic response is an important method, but the human access to examining the pipe condition is significantly limited because these pipes are usually placed in a very confined space, where even replacing a battery for sensors by hand is difficult. Self-powered remote sensing is thus desired to monitor the condition of the pipes, especially for long-term operation. Therefore, energy harvesting from a local source is necessary. A set of hot and cold pipes next to each other could be an energy source for thermoelectric energy harvesting. In this section, an analytical design optimization of thermal energy harvesting with a thermoelectric module is conducted, followed by the experimental validation.

Advanced sensors and wireless transmitters consume power only in a range of 1–10 mW for their minimum intermittent operation. For temperature differences of 1–10°C, we investigate two different heat transport approaches: (1) heat pipes and (2) mini-channel cold plates, both of which can fit a confined 0.1 m gap space. In experimentation, purified water runs through the hot and cold pipes and the temperatures are controlled by two independent scientific chillers. The pipe diameter is significantly smaller compared to the actual one, but the gap between the pipes is maintained. The results shown are reasonably well matched with the analytic model, while the TE modules were not optimally designed. A series of an off-the-shelf TE modules with a size of 40 mm × 40 mm and containing 127 thermocouples is used. The electrical power harvested from the set of 30°C and 7°C water are found to be 2.4 mW and 81 mW for the heat pipe and mini channel, respectively. The output voltage at the optimum load was slightly lower than 1.0 V. If it is designed to reach 1.2 V, a DC−DC converter will be applicable for 5 V universal serial bus (USB) interface, which widely matches a variety of off-the-shelf monitoring equipment.

7.1.1 Case study

In general, buildings have many pipelines that transport heat by water, steam, or some other refrigerants. Also, datacenters require a lot of chilled water supply [1, 2]. These pipes are likely placed in parallel in a confined space, as shown in Fig. 7.1.

It is often the case that the two fluids in these parallel pipes are at two different temperatures. This could be an energy source for powering equipment used for sensing and monitoring the temperature, flow rate, acoustic noise level, and so on. Monitoring these conditions are particularly important to identify the healthiness of the pipelines [4]. Due to the expensive footprint for such highly intelligent and/or energy-intensive buildings, the pipes are constructed in a very confined space, so that a person would have difficulty reaching them even for reading data or replacing the battery. Hence, an unmanned operation is desired. In addition, wireless networking of sensors has been investigated [5]. On-spot thermoelectric (TE) power generators are gaining attention as a promising technological alternative to replace batteries for such maintenance-free power supply for the sensors and wireless transmitters, so that human access to the sensors would be no longer necessary. This section focuses on the energy harvesting from two different temperature parallel pipelines.

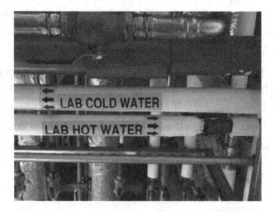

Fig. 7.1: Example of pipelines in a building: A picture of piping in Birck Nanotechnology Center at Purdue University [3].

7.1.2   Thermal design

Heat transfer design determines the generator capacity using a fraction of heat energy by using the temperature difference between two fluid flows. There are essentially two approaches. One is to make a bypass of heat flow by heat conduction and another is to make an additional path of fluid flow.

7.1.2.1   Type-1: Heat flow bypass

Two identical and small diameter heat pipes are used to conduct the heat from the pipelines to the hot side and cold side conductive plates with thermoelectric. Heat pipe transports the heat using two-phase cycle between two terminal temperatures. Effective thermal conductivity $\beta_{eff} = QL/(\Delta T A)$ [W/(m K)] of the heat pipe is much larger than a copper rod. The effective thermal conductivity is found by characterizing all right-hand-side parameters of its formula. It is estimated at 1500 W/(m K). This value is relatively conservative compared to the literature [6, 7] for larger ΔT electronics cooling applications. The detailed description of the heat pipe mechanism can be found elsewhere.

Although the following experiment used a cylindrical heat pipe, advanced flat heat pipes [8] are popular in the market and could be utilized in thin profile for this implementation. A 1 mm thick flat heat pipe could transport more than 10 W per unit, for example. The heat pipe was expected to transport the heat within a temperature difference of 5°C in this application.

The heat pipe is inserted into a copper plate, which is directly attached to the thermoelectric power generator (TEG) module with a thermal grease (Fig. 7.2). There are three major thermal resistances involved, the heat pipe itself ψ_{hp}, spreading thermal resistance in the copper ψ_{sp}, and the contact interface thermal resistance ψ_{if}. These are determined as follows:

$$\psi_{hp} = L/(\beta_{eff} A) \qquad (7.1)$$

$$\psi_{sp} = \frac{\lambda}{\beta_s \alpha (1 + 2\lambda tan\phi)} \qquad (7.2)$$

$$\psi_{if} = 0.15 \qquad (7.3)$$

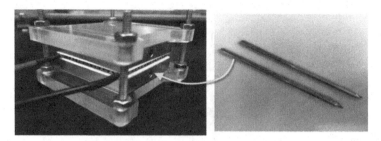

Fig. 7.2: TE generator locates in between the copper plates, bolted in a couple of polycarbonate brackets. Heat pipes (right, φ 3 mm, provided by Fujikura Ltd.) are inserted into the holes in the copper.

Total thermal resistance from the fluid to the TEG contacts are found as $\psi_{total} = 7.44$ [K/W]. The effective heat transfer coefficient h_{eff} at the contact area of 0.04 m $\times$ 0.04 m is found to be 84 W/(m² K).

7.1.2.2 *Type-2: Water flow bypass*

Flow bypass approach is based on making an alternate path with a fraction of water flow for utilizing TEG with a mini-channel cold plate for both hot and cold sides. In practice, pipelines need to allow some modification for the flow bypass, but eventually this approach has been found to be very effective. The cold plate, in this particular device, consists of 44 interim channels in the mid-section manufactured by aluminum extrusion technology, note however that only 36 channels are directly involved in heat transfer due to the dimension mismatch (Fig. 7.3). The hydraulic diameter D_h is 1.15 mm. The channels are long enough to consider a fully developed flow. Within laminar flow range, Nusselt number is 2.44 [9] with one side heated and remaining three sides being adiabatic at fixed temperature boundary condition. Then, effective heat transfer coefficient at the contact surface is found to be 2400 W/(m² K):

$$h_{eff} = \frac{A_{eff}}{A_{wett}} \frac{\beta_f}{D_h} \text{Nu} \qquad (7.4)$$

7.1.3 *Experimental apparatus*

The setup mimics a couple of parallel water pipelines with 10 cm distance in between. Two water flows are generated by lab scale

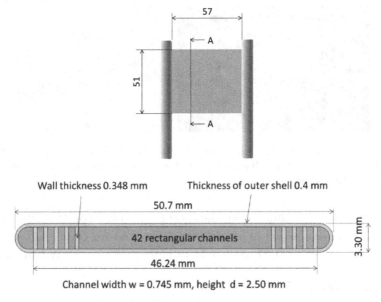

Fig. 7.3: Miniature aluminum cold plate with parallel mini channels inside (CP-20, Lytron). The internal channel dimensions in the cross-section (A–A) are values measured by physically cutting the part.

identical chillers, where inlet flow temperatures are controlled in ±0.1°C accuracy (ISOTEMP 250LCU, Fisher Scientific). The water mass flow is manually controlled with the valves. Test section has 3/8-inch standard (9.5 mm I.D.) pipes. The flow meters measure the volumetric flow rate $\dot{V}$ [mL/min] by a turbine method. Two identical miniature flow meters are used for each flow (FTB333D, Omega Engineering) with ±6% accuracy for 0.2−2 mL/min. TEG modules are placed between copper plates with good thermal interface with grease (OMEGA THERM201, Omega Engineering) and the test section is insulated to minimize heat leaks. In the power output measurement, load resistor is chosen from the available resistor. For CP14-127-045 module, 2 Ω is used while the model predicts 2.1 Ω. Similarly, for CP14-127-10 module, 3.9 Ω is used while 3.32 Ω is predicted. Except for a different test section of Type-1 and Type-2, the entire arrangement is as shown in the schematic diagram (Figs. 7.4–7.6).

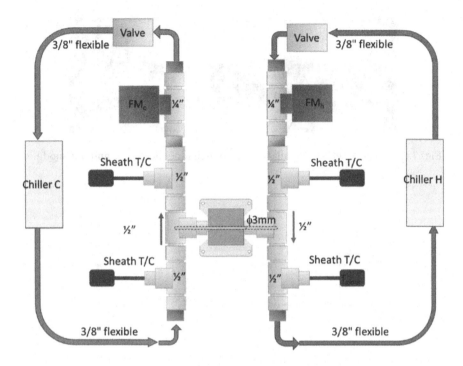

Fig. 7.4: Schematic diagram of the test setup.

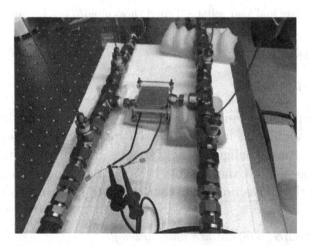

Fig. 7.5: Type-1 setup. Fluid-thermal test section. The block in the middle of the pipes is the test section with a TE module. The output leads are connected to the load resistor (1/4 W class resistor) and the voltage meter.

Fig. 7.6: Type-2 setup. Installation of mini-channel heat sink and TE module.

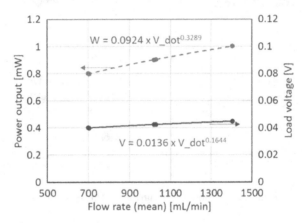

Fig. 7.7: Power and load voltage. Type-1 (heat pipe), $T_a = 7°C$ and $T_s = 30°C$.

7.1.4 *Analysis*

7.1.4.1 *Flow rate*

Power output as a function of flow rate is generally moderate. The impacts are different for Type-1 and Type-2 (Figs. 7.7 and 7.8). For the flow bypass configuration (Type-2), gradual saturation of the curve suggests that the sensible heat by mass flow approaches the limit and the convection heat transfer drives more in this condition.

7.1.4.2 *Source temperature and leg length*

The maximum power output is in general proportional to the square of temperature difference. The impact of leg length is relatively rarely

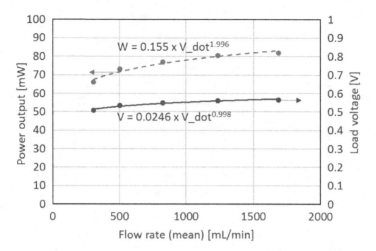

Fig. 7.8: Power and load voltage. Type-2 (cold plate), $T_a = 7°$C and $T_s = 30°$C.

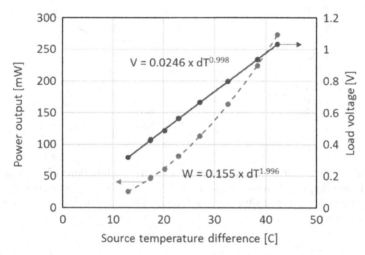

Fig. 7.9: Power and load voltage as functions of DT of pipelines. CP14-127-10 module (longer leg) with Type-2 (mini-channel heat sink) and flow rate of 1.7 l/min.

studied due to the challenge in scaling up the TE leg length in manufacturing. As discussed in the earlier section, there is also a significant benefit for an optimally designed leg, as seen in Figs. 7.9 and 7.10.

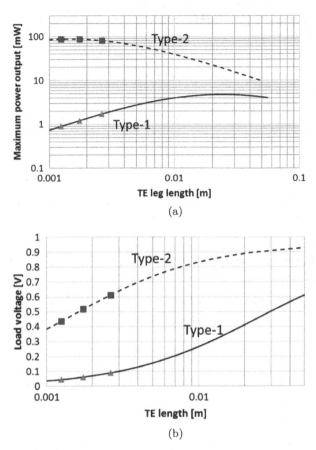

Fig. 7.10: (a) Power output and (b) load voltage as functions of leg length for
Type-1 and Type-2. Curves represent the model and dots represent the data
from experiments. Three different leg lengths were tested: 1.24 mm, 1.74 mm,
and 2.65 mm.

The Type-1 design with moderate heat transfer of 84 W/(m^2 K)
requires an order of magnitude longer (thicker) leg compared to the
current off-the-shelf modules to achieve maximum power output.
The Type-2 design with heat transfer of 1519 W/(m^2 K) is quite different. Off-the-shelf design of legs more or less matches the optimum,
while the theoretical optimum is a length of 2.65 mm. Peak voltage at

the load of 0.6 V is still lower than required to drive a USB interface (5 V) electronics.

7.1.4.3 *Energy efficiency*

Energy efficiency is meaningless for this case since only a very small fraction of the energy is used. However, power output per unit module matters in terms of cost investment for the device. With a condition of 30°C to 7°C ($\Delta T = 23$°C), as an example, energy conversion efficiency is 0.86% with Type-2, implying that heat energy drawn from the hot side is around 10 W for 86 mW of power output.

7.1.4.4 *Accuracy*

The experimental results reasonably matched the analytic model discussed in the earlier sections. The predicted power output based on the measured geometry and assumption of the material properties are relatively lower than the experiments with from -17% to 0% error.

7.1.4.5 *Alternative ideal design*

To drive a 5 V DC electronics, two design modifications will be the options for the case of $\Delta T = 27$°C condition with Type-2. One is to enlarge TEG to increase the series number of contacts of the Seebeck effect, or another is to add a DC–DC converter.

The TEG module including the leg shape is supposed to be designed from scratch with current materials and process technologies. Module size should be enlarged from 40 mm × 40 mm (127 couples) to 50 mm × 50 mm (197 couples) to extend the load voltage. With a leg length of 3.35 mm, the load voltage is predicted to hit 1.2 V. With this design, the power output of the module is then 56.9 mW, while the ideal peak power is 59.6 mW at 1.13 V. By implementing a DC–DC converter with an efficiency of 80%, output is predicted to be 45.5 mW at 5 V DC (Figs. 7.11 and 7.12).

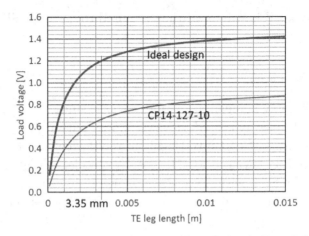

Fig. 7.11: Load voltage vs TE leg length for off-the-shelf module and ideal design. To achieve 1.2 V of load voltage, 3.35 mm of leg length is required for the ideal design of module.

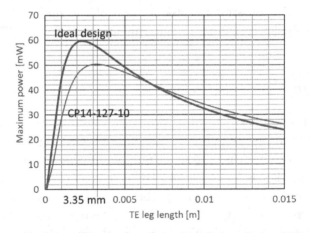

Fig. 7.12: Module power output vs TE leg length for off-the-shelf module and ideal design. The leg length to match 1.2 V is slightly off the maximum power output.

7.2 Human Body Heat Recovery

In this section, the performance of energy harvesting from a human body heat, specifically from arms, with a thermoelectric energy conversion is discussed. Wearable heat recovery devices are expected to supply electricity to a few embedded or wearable sensors as well as a wireless transmitter circuitry, so that no battery is needed. Currently, wireless transmission of monitoring data, such as heartbeats and body temperature, facilitates accurate and time-sensitive healthcare. Thermoelectric helps this new approach with recovery heat to drive electronics as it thermally makes contact with the body surface.

There is a limitation of power due to the low heat generation rate by the biological system nature. Maximum power harvesting is critical without creating discomfort. The energy dissipation from the skin, however, is highly dynamic and wide-ranging, depending on the human activities, such as staying quiet, walking, sporting, or exercising. According to [10], in stationary mode, a human dissipates around 40 W/m^2 and it increases up to an order of magnitude larger than 640 W/m^2 when someone is engaged in vigorous exercises.

7.2.1 Review of previous study

In this section, the study on thermoelectric power generation from body heat is reviewed. Thermoelectric was considered as a cooling device [11, 12] early on in the 1960s for medical purposes. Some of the thermoelectric applications also helped to cool the human body under extreme conditions [13]. However, a power generator using the human body came out as a practical working device a few decades later.

Wristwatches driven by body heat were commercialized [14, 15] in 1998 and 2001, respectively. Since the wristwatch needs to continuously operate even without body heat, such a product contains a parallel battery. Hence, it is not very clear how much percentage of power driving the system was harvested from body heat. Leonov and his group at IMEC had pioneered with their work on harvesting body heat with thermoelectric devices to drive electronics in late 1990s and early 2000s. They investigated and developed wristband-type power generators [16, 17]. In a more recent approach, they used a stack thermoelectric module with a fin-type heat sink attachment. Utilization of thicker elements for this low heat flux application is important for obtaining a useful temperature difference, and the bulky heat fins obviously represent the need for heat transfer enhancement on the cold side. While a thermal design match for the application is important, the electric circuitry for practically harvesting the energy is also a very important part. The thermoelectric module design particularly for wristband application and the orientation impact were reported in Ref. [18].

Harb [20] studied various energy harvesting systems from the human body as an energy source, including the thermoelectric power generator. Harvesting momentum energy with piezoelectricity may be useful along with steady thermal energy with thermoelectricity. Chandrakasan *et al.* [21] discussed the ultra-low voltage operational device and application along with integration of energy harvesting devices. Considering that a typical utilization of wearable energy devices is in small area and the energy density out of the human body is very low, the low voltage output must be a key component for the design of thermoelectric generators. Voltage regulation is then an important requirement for electronic applications. Wireless electronics are quite popular nowadays and the role of communication with embedded sensors is becoming more important. Vullers *et al.* summarized the wireless sensors and their local power sourcing [22]. The seamless power connection to the wireless sensor may create a significant benefit for providing a local power generator to harvest the human body heat. Mateu *et al.* [23] also investigated the wireless sensor matched energy harvesting systems. Hoang *et al.* [24] investigated powering medical healthcare devices with body heat. Watkins *et al.* [25] from the RTI group reported on a thermoelectric module aiming to provide power for embedded medical devices, e.g. pacemakers using power from the body heat of the skin. This approach seems to be setting a somewhat miscalculated target. Generating 100 μW from 6.8 cm^2 of size from human skin may require about 0.37% efficiency and about 7.3°C of temperature difference across the element, while the target temperature difference is up to 1.7°C, which sounds reasonable. Concerning flexible body heat harvesting, Jo *et al.* [26] experimentally studied the polymer-based thermoelectric module and reported 2.1 μW of power output. Kim *et al.* [27] investigated a unique approach using glass fabric as the substrate to obtain a flexibility for the thermoelectric module. This does not necessarily fit to a body heat recovery application, but Stark [28] made a useful packaging of thermoelectric generators similar to wearable batteries.

From the materials point of view, poly(3,4-ethylenedioxythiophene): polystyrene sulfonate (PEDOT:PSS) has received a lot of attention as an efficient organic thermoelectric material due to its higher electrical conductivity and controllability of doping level and mechanical flexibility. Also, its printable process has great potential for future high-volume production. Park *et al.* [29] investigated polymer thermoelectric materials for generating power by the touch of fingertips. There are excellent recent reviews on organic thermoelectric materials [30–32] for flexible and wearable thermoelectric energy conversion.

7.2.2 *Optimum design for body heat recovery*

While there are unique considerations when dealing with body heat recovery (such as a small temperature gradient and low heat flux), the objective of designing thermoelectric device is maximizing the harvested power.

The heat flux to the TEG is denoted as q [W/m^2], which represents the heat supplied by a blood flow. The blood temperature inside the body is maintained at around 37°C by the biothermal system of human body. Therefore, the heat source is considered as a temperature reservoir. The heat flows through the circuit down to the ambient air. The ambient air has a significantly larger capacity, so it is considered as a temperature reservoir as well.

In an ideal design for the maximum power output, the leg thickness d does not have much flexibility due to the thermal conductivity, the fill factor and the heat flux limits. The optimum design may likely go beyond 1 cm thick for many cases. In practice, the design of the module with off optimum may be unavoidable. In the static mode, such as desk work consisting of sitting on a chair, the heat generation from the blood supply is estimated around 40 W/m^2, which differs between 30 W/m^2 [33] and 60 W/m^2 [34]. In the dynamic mode, such as playing a motion-intensive sport or doing hard exercise, the heat generation as the dissipation of muscular energy consumption could reach 600–700 W/m^2.

7.2.3   *Parasitic loss consideration*

Parasitic losses include the following: (1) thermal bypass, (2) thermal contraction/spreading, (3) electrical contact, and (4) electrical intrinsic resistance. The fill factor plays a significant role as the area coverage is getting lower, e.g. $F < 20\%$. Details of the thermal and electrical components of parasitic losses are discussed in Chapter 5.

  Figure 7.13 shows the result of a case study for body heat recovery. Note that the power output is significantly small when

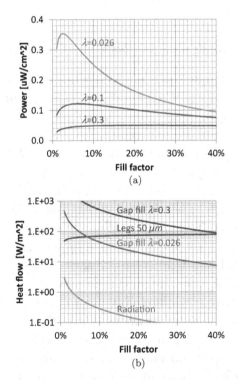

Fig. 7.13:   (a) Power output vs fill factor in variations of thermal conductivity of gap fill materials, 0.3, 0.1, and 0.026 W/(m K). Leg thickness is fixed to 50 μm. (b) Heat flow vs fill factor, with gap fill materials, 0.3 and 0.026 W/(m K). Leg thickness is optimized. Radiation heat transfer contribution is overplotted. Temperature conditions for both are $T_s = 32°$C and $T_a = 24°$C.

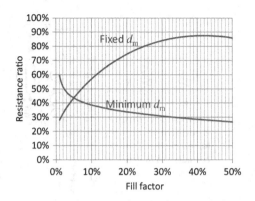

Fig. 7.14:   Electrical resistances ratio vs fill factor.

the thermal conductivity of the gap fill is similar to that of the thermoelectric leg.

Figure 7.14 shows a plot of electrical resistance ratio (intrinsic series resistance against total resistance) assuming 5 μm of thickness for electrode and $10^{-7} \Omega \, cm^2$ for the specific contact resistance value, which is known as a good practice for the conventional bismuth-telluride-based modules. Note that the minimum electrode thickness d_m is 5 μm at a 5% fill factor. Hence, the fixed thickness is assumed to be 5 μm.

7.2.4   *Transient response and analysis*

In order to understand the transient response of the module when it makes contact with skin, transient thermal model is developed to analyze the temporal behavior.

The analytic model of TEG is extended by introducing thermal capacitors in the circuit, as shown in Fig. 7.15. The thermal capacitors contain the thermal mass of skin and the TEG module itself. In this case, (1) the thermal contact involves almost zero thermal mass and (2) convection already exists on the body surface, so both impacts are negligible. The electrical time constant is much smaller than the thermal response enough to be omitted in modeling the thermal transient response.

To obtain the temperatures as a function of time, a quasi-steady state model is used in our numerical approach. Thermal diffusion

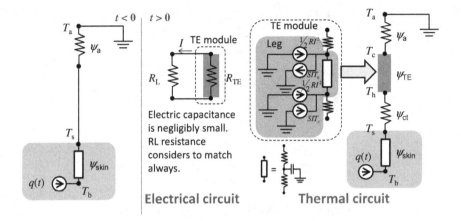

Fig. 7.15: Transient thermal and electrical circuits.

is solved to determine the temperatures at the two contacts of thermoelement first. Then, the temporal power output is calculated using the two temperatures at any time point. A 1D thermal diffusion equation is described as

$$\frac{\partial T}{\partial t} = \alpha \frac{\partial^2 T}{\partial x^2} \tag{7.5}$$

where $\alpha = \beta/\rho C$ is diffusivity [m²/s] with density ρ [kg/m³] and specific heat C [J/(kg K)], x [m] is the distance from the heat contact, and t [s] is the elapsed time since the contact occurred ($t = 0$).

7.2.4.1 *Numerical discretization of thermal diffusion*

For sufficiently small Δt and Δx, Eq. (7.5) can be replaced as a linearized equation as

$$\frac{T_{x,t+\Delta t} - T_{x,t}}{\Delta t} = \alpha \frac{T_{x-\Delta x,t} - 2T_{x,t} + T_{x+\Delta x,t}}{(\Delta x)^2} \tag{7.6}$$

Rewriting the equation to formulate a matrix for the calculation with i representing Δx and j representing Δt,

$$T_{i,j} = \frac{\alpha \Delta t}{(\Delta x)^2} \left\{ T_{i-1,j-1} + \left(\frac{(\Delta x)^2}{\alpha \Delta t} - 2 \right) T_{i,j-1} + T_{i+1,j-1} \right\} \tag{7.7}$$

To solve the equations, boundary conditions are applied at the convection side as

$$T_{0,j+1} = \frac{T_a + (1/\text{Bi})\,T_{i+1,j+1}}{1 + (1\text{Bi})} \text{ with Bi} = \frac{h\Delta x}{\beta_1} \qquad (7.8)$$

where Bi is called the Biot number where β_1 is thermal conductivity of layer-1 (the surface layer contacting the open-air convection) and h is heat transfer coefficient [W/(m^2 K)] at the surface. Also, at the heat source contact (the bottom layer contacting blood flow) $i = n$,

$$T_{n,j+1} = T_{n-1,j+1} + q\frac{\Delta x}{\beta_n} \qquad (7.9)$$

where q [W/m^2] is the heat generation per unit area supplied by blood flow and β_n is the thermal conductivity of layer-n.

7.2.4.2 *Multi-layer with contact resistance*

In order to investigate the contact between the skin surface and the thermoelectric module, the following model is used. Figure 7.16 shows the insertion of the lumped contact resistance, which is considered as no thermal mass.

Instantaneous energy balance at both sides of the contacts is

$$\frac{\beta_1}{\Delta x_1}\left(T_{k,j} - T_{k-1,j}\right) = \frac{1}{\psi_{ct}}\left(T_{k+1,j} - T_{k,j}\right) = \frac{\beta_2}{\Delta x_2}\left(T_{k+2,j} - T_{k+1,j}\right)$$

$$(7.10)$$

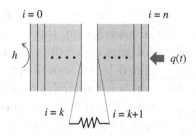

Fig. 7.16: Multi-layer model with a lumped contact resistance without capacitance.

Applying the above to the model, the temperature at both sides of the contacts is described as

$$T_{k,j} = \frac{1}{(\kappa_1 + 1/\psi_{ct})}\left(\kappa_1 T_{k-1,j} + T_{k+1,j}/\psi_{ct}\right)$$

$$T_{k+1,j} = \frac{1}{(\kappa_2 + 1/\psi_{ct})}\left(\kappa_2 T_{k+2,j} + T_{k,j}/\psi_{ct}\right) \tag{7.11}$$

where κ_i $(i = 1, 2)$ is the thermal conductance across the layer with thickness of Δx

$$\kappa_i = \frac{\beta_i}{\Delta x} \tag{7.12}$$

In the time progression,

$$T_{k,j+1} = r_m\left(T_{k-1,j} - 2T_{k,j} + T_{k+1,j}\right) + T_{k,j} \tag{7.13}$$

where r_m is the material-dependent diffusion factor defined as

$$r_m = \frac{\alpha_m \Delta t}{(\Delta x_m)^2} \tag{7.14}$$

The transient thermal profile is shown in Fig. 7.17. The power output is based on these thermal profiles. Figure 7.18 shows the power output according to the device properties and the dimensions: thermal conductivity 0.3 W/(m K), Power factor 9×10^{-4} (m K^2)/W, where $ZT \sim 1$, as well as the leg length 0.05 mm and fill factor 0.2. The heat flux across the TEG device is also shown in the graph.

7.2.5   *Stationary and dynamic heat generation*

Power output is a function of heat rate through the conversion device and its efficiency. The definitions of stationary mode and dynamic mode are similar to the evaluation metric of a fuel economy in automotive vehicles [35]. In the near future, the evaluation criteria may need to be standardized for wearable power device products. Here, the stationary mode is defined by no movement of body, hence the internal energy consumption is minimum and mostly used for maintaining the body temperature. The stationary mode is not affected by forced convection heat transfer but is affected by passive air convection. Dynamic mode, in contrast, is defined by the condition of dissipating heat from the mechanical energy consumption by the muscles.

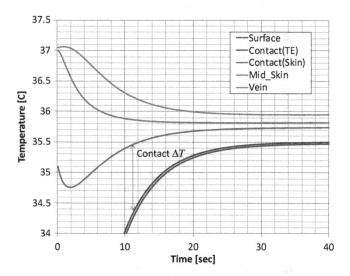

Fig. 7.17: Thermal time response. Conditions: Constant uniform heat supply 40 W/cm^2 at skin with constant blood temperature of 37°C. Ambient air temperature: 23.5°C. Skin: thickness 2 mm, thermal conductivity 0.37 W/(m K), density 1109 kg/m^3, specific heat 3391 J/(kg K), contact thermal resistance 65 K cm^2/W. Total thickness of module 400 μm, effective thermal diffusivities of the thermoelectric module and the skin are 2.75 × 10^{-7} m^2/s and 9.84 × 10^{-8} m^2/s, respectively. At the steady state, temperature difference across the TE element is 0.028 K.

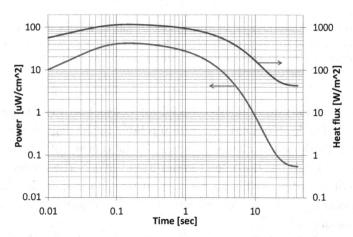

Fig. 7.18: Transient power generation and the heat flux through the TEG with conditions and properties the same as in Fig. 7.17.

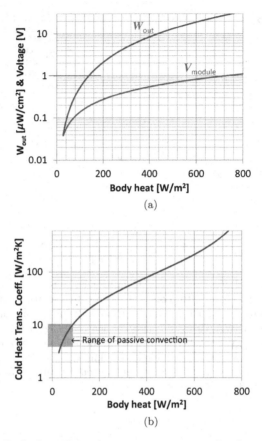

Fig. 7.19: (a) Body heat-dependent power output and voltage for a 100 cm^2 area module with the element density 1 leg per 1 mm × 1 mm footprint with fill factor 10%. (b) The cold side heat transfer coefficient required for the power output at the same conditions.

Due to the body movements, forced air convection heat transfer is involved. The level of heat generation relates to the degree of motion. Figure 7.19 shows an ideal case with (1) the power output and voltage as a function of body heat generation and (2) the required cold side heat transfer coefficient for the future device spec. During the time period, body surface temperature is maintained at a constant temperature of 35°C.

As seen in Fig. 7.19(a), the power output can drastically increase as the heat generation rate increases. This is because the heat flow across the thermoelectric module increases. However, this also means that the effective heat transfer coefficient (Fig. 7.19(b)) from the surface is required to be very large at the same time to maintain the skin temperature.

7.2.6 *Potential enhancement*

In target of 1 μW/cm^2 of power output for wearable applications, the following summarizes the potential design improvement for each component.

(1) *Thermoelectric leg*: Utilizing TE legs thicker than a few millimeters usually may yield a better performance for body heat recovery application. The fill factor is an alternative parameter to adjust to make the design closer to optimum. Typically, a low fill factor less than 20% is desirable. Due to the relatively small electrical resistances of the TE legs, it is also important to reduce the contact resistance.

(2) *Gap fill material*: Lower thermal conductivity of the gap fill relative to the thermoelectric leg is highly desired. Air gap is acceptable, but it allows parasitic radiative heat loss through the gaps. Hence, materials such as aerogels may be a reasonable solution.

(3) *Electrode layers*: In order to reduce electric parasitic losses, consideration of the thickness of the electrode layers is also important. At least a few tens of microns are expected.

(4) *Heat dissipation surface*: Fins or surface area extension works under forced convection condition. Fins shorter than the boundary layer may not be very effective, especially for a thicker boundary layer of natural convection.

(5) *Voltage converter and load impedance*: As discussed in Section 6.1, DC–DC conversion, even with sacrificing 10–20% of power, is available only the case natural TEG load output exceeds 1.2 V.

7.2.7 *Summary of body heat recovery*

Based on the generic TEG model, the unique condition of human body heat recovery was modeled. Specifically, available low heat flux and small temperature difference allow a small power output. Also, the leg thickness match the maximum power output is quite large, while the design with a reasonable thickness for flexibility finds it hard to obtain the maximum. However, based on the model, there are several approaches that lead to the enhancement of the power output. Transient modeling allows one to predict not only the time response of power output but also the temperature change on the skin surface. In future, this could be a route guidance to explore the comfortability of such devices.

7.3   Flexible Thermoelectric Generators

From an application point of view, having a 'wearable' feature for electronics is a big advantage. Wearable devices are expected to be flexible enough to be conformally attached to human skin-like fabrics. The degree of flexibility is a matter of application, but essentially these devices must be repeatedly bendable and cannot be easily separated from the target surface during the wearer's movement. Thermoelectric materials used may not be necessarily flexible, but the generator device needs to be flexible. A comprehensive review of wearable thermoelectric generators has been presented in the literature [36]. Figure 7.20 shows an application example of wearable generators from the review.

7.3.1   *Polymer-based thermoelectric*

Polymer-based thermoelectric materials are of great interest because of the flexible nature of the materials, which come from the structure of large molecules. The binding force between the molecules are relatively weak in polymers, which also implies that the phonon energy transmission across the material may be weak. In fact, the thermal conductivity of polymeric materials is nearly an order of magnitude smaller than the rigid semiconductor thermoelectric materials. Polymer-based materials are also suitable for mass production process using, for example, printing technologies, so that the raw material cost of manufacturing thermoelectric device is likely lower than conventional semiconductor devices. The unique characteristics are

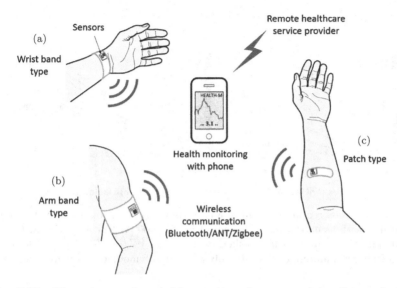

Fig. 7.20: Three types of wearable sensor nodes powered by thermoelectric energy harvesters. The thermoelectric generators are preferably made of flexible materials and substrates, so that they can be conformally attached onto the various locations of the skin with enhanced thermal contact. Monitored data are transmitted via a short-distance wireless communication protocol such as Bluetooth, ANT, or Zigbee to a portable personal server such as a cell phone and then to the remote healthcare service provider via a long-distance network.

Source: Reproduced with permission from *J. Mater. Chem. C*, **3**(40), 10362–10374 (2015).

analyzed compared to semiconductor materials in Ref. [37]. According to the investigation, polymer-based materials in general are better suited for low heat flux applications. In Figs. 7.21 and 7.22, the required mass for unit power output (kg/W) and the power cost ($/W), respectively, are plotted with variations of figure of merit (ZT) and fill factor. Two different heat transfer coefficients at the boundary of thermoelectric leg, 20 kW/(m^2 K) (left plots), and 10 W/(m^2 K) (right plots), representing the high and low heat flux regimes, respectively, were selected and compared to reflect different heat source capacities and heat sink performances.

Despite the cost merit and the flexibility benefit, some noticeable concerns remain for polymer-based materials. A gap filler is required for laminate process fabrication or the structural integrity of the

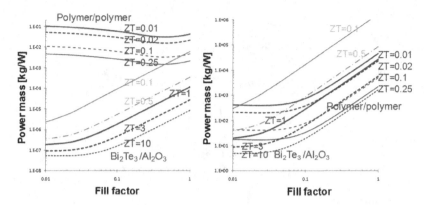

Fig. 7.21: Mass required for unit power output [kg/W] vs fill factor at (left) high heat flux with heat transfer coefficient of $h = 20,000$ [W/(m^2 K)] and low (right) heat flux with heat transfer coefficient of $h = 10$ [W/(m^2 K)]. A module made from BiTe with alumina substrate polymer and polymer substrate are compared.

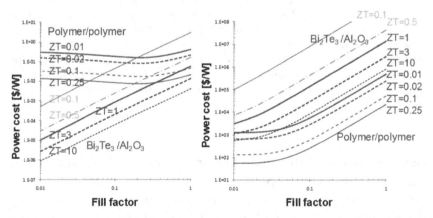

Fig. 7.22: Power cost [\$/W] vs fill factor at (left) high heat flux with heat transfer coefficient of $h = 20,000$ [W/(m^2 K)] and low (right) heat flux with heat transfer coefficient of $h = 10$ [W/(m^2 K)]. A module made from Bi$_2$Te$_3$ with alumina substrate polymer and polymer substrate are compared.

device. Thermal conductivity of the gap fill material is desired to be much lower than that of the thermoelectric polymer, which may not be easy.

PEDOT:PSS is a popular organic material that many researches have been exploring for body heat energy harvesting. This material was originally developed as an electrically semi- or true-conductive polymer and was widely controlled by doping. It has been considered for semi-transparent electrode and utilized for organic solar photovoltaics [38].

In addition to the mechanical flexibility, lower thermal conductivity of polymer material is a big advantage, where most of the thermal conductivity of polymer-based thermoelectric materials falls into the range 0.2–0.5 W/(m K), an order of magnitude smaller than typical semiconductor solid thermoelectric materials. This suggests a range of an order of magnitude thinner thermoelectric leg matches with the system compared to the semiconductor materials.

7.3.2 Folded film thermoelectric

Thermoelectric materials in some cases are milled into small particles and then consolidated into a solid pack or directly into legs by spark plasma sintering (SPS) or similar high-pressure and high-temperature processes. This process is effective for enhancing the thermoelectric performance of the material, since the grain boundary creates additional phonon scattering to reduce the thermal conductivity, with relatively unaffected electron transport.

The small grain particles can be a main material of ink for painting or printing processes with some binders added as long as the size of particles is well controlled. The thermoelectric materials and metal pattern are printed on a film. Then, a module is constructed by stacking multi-layers and electrical connection is made afterwards. Another method is to fold the printed pattern many times to increase the cross-sectional area until it reaches a realistic size. A German company OTEGO, spin-off from the Karlsruhe Institute of Technology, developed an automated process of manufacturing folded thermoelectric modules [39], which has moderate mechanical flexibility to fit to a cylindrical surface, such as straight pipes [40].

7.3.3 Woven thermoelectric

One new concept is a woven fabric thermoelectric module made from weaving thermoelectric strings which have a repeating [(p-type)-metal−(n-type)-metal] (p-m−n-m) pattern with bismuth telluride (Bi_2Te_3) together with insulation yarns. This section shows the calculated performance and cost estimation of the module compared with a conventional (Π-structure) module that consists of a polymer-based thermoelectric material such as PEDOT:PSS.

7.3.3.1 *Challenge in design of flexible thermoelectric modules*

Flexible thermoelectric modules are actively being researched. Most of the work is based on organic thermoelectric materials. One advantage is that organic material has naturally low thermal conductivity, and the electric conductivity can be modified drastically. The recent lab scale performance reported was in the $ZT \sim 0.5$ range [41, 42]. Hence, there is a challenge in manufacturing such flexible thermoelectric legs in an array which is the same as that of a conventional thermoelectric module. Automated pick-and-place fabrication is no longer adaptable for non-rigid legs. And soldering at higher temperatures of 230–270°C [43] is no longer available for making electrical contact. This manufacturing concern is one of the barriers to transforming the technology for industrialization.

7.3.3.2 *Performance of woven modules*

Performance calculation is based on human body heat recovery. The skin surface is assumed to maintain a temperature of 35°C, while the ambient temperature is 25°C. These boundary conditions are fixed, but the temperature contacts of the thermoelectric are dependent on thermal design and heat transfer. The contact effective heat transfer coefficient is 153 W/(m K) based on the fingertip contact example [44]. The heat transfer coefficient for the air-cooling side is 4 W/(m K) based on the 40 W/m^2 heat dissipation in static mode. This is nearly the same as the correlation for the natural convection along the vertically oriented wall considering the characteristic length if the wall is in the order of 10 cm.

The material properties of the thermoelectric materials include a thermal conductivity of 0.5 W/(m K), an electrical conductivity of 1.27×10^5 1/(Ω m), and a Seebeck coefficient of 80 μV/K. The thermal conductivity of the laminate films is 0.16 W/(m K) and the gap is filled by air 0.026 W/(m K). The specific contact resistance

10^{-5} Ω cm^2 assumes one order of magnitude larger than an ordinal contact resistance.

The model is based on the conventional Π-configuration including thermal and electrical parasitic losses with thermal spreading and electrical contact and series resistances. The change needed for this woven configuration is only one point for the parallel thermal conductive heat loss through the gap between the two film laminates (substrates). To calculate the correct thermal conductance of the gap, whether the gap is filled in or not, the conductivity of the gap fill material virtually increases by the ratio of leg length over gap height. Fortunately, thermal conduction in a lateral direction of fabric is essentially zero because the heat conduction across the legs is one dimensional from one side to another.

7.3.3.3   *Calculation*

Assumptions made for this calculation are as follows:

1. Number of legs: 100 per 1 cm^2.
2. The gap filled with air and radiation thermal cross-talk between the laminates is negligible (due to the very small temperature differences). The air gap has $1/10$–$1/50$ smaller thermal conductivity compared to the thermoelement.
3. The electrical contact is similar to that of the ordinal thermoelectric modules, but the series resistance is extremely small due to the string structure.
4. The thickness of the outer shell film laminates is 70 μm each (Fig. 7.23).

The effective fill factor F is found as

$$F = \frac{x\,(L/d)}{2p_w p_y}d^2 \qquad (7.15)$$

where x is the fractional contact area, L is the actual leg length, d is the string diameter, p_w is the half-pitch between the contacts on a film substrate along the leg length direction, and p_y is the warp thread pitch.

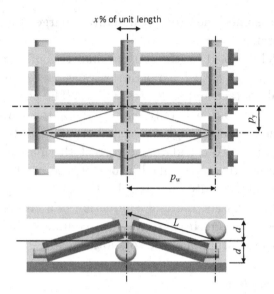

Fig. 7.23: Geometry and dimensions of the woven structure. An effective area for a contact is a lozenge area emphasized by the diamond shape. Good thermal contacts between the metal interconnects and the top and bottom heat transfer surfaces, respectively, are a key requirement for optimum operation. Weft yarns are low thermal conductivity material.

Figures 7.24 and 7.25 show the power output (μW/cm^2) and power cost ($\$/\mu$W), respectively, in variations of the weaving string diameter. Since the above parameters are functions of the woven ratio, L/d, the results vary. In this calculation, metric of flexibility is not calculated. As a reasonable statement, weaving with larger diameter is likely to be less flexible. The parameters used in these calculations are 35°C of body temperature and 15°C of ambient temperature. The effective heat transfer coefficient for the skin contact is 154 W/(m K) and the air convection is 6 W/(m K). The L/d ratio is always optimally designed, so that effective fill factor accordingly varies (0.017–0.040). Thermoelectric material is assumed to be Bi$_2$Te$_3$ with $ZT = 1$ with the thermal conductivity of 2.0 W/(m K). Both sides of the module have 127 μm thick laminates made of polyimide. The densities are 8,900 kg/m^3 and 1,300 kg/m^3 for thermoelectric material and laminate tapes, respectively. The raw material costs are 500 $\$/$kg, 100 $\$/$kg, and 29.2 $\$/$m^2 for thermoelectric material, core fiber, and laminate tapes, respectively.

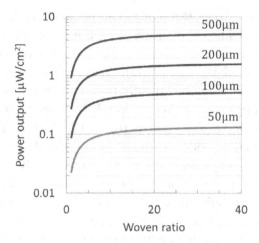

Fig. 7.24: Power output per unit area $[\mu\text{W}/\text{cm}^2]$ in variations of string diameter from 50 μm to 500 μm.

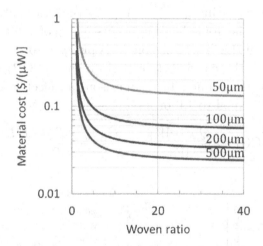

Fig. 7.25: Material cost per unit power output $[\$/\mu\text{W}]$ in variations of string diameter from 50 μm to 500 μm.

7.4 Thermoelectric Thermotherapy

This section highlights an application of heat pumping mode using thermoelectric, especially for low heat flux and small temperature change. In order to provide a medical care for muscle, thermoelectric can be used as a thermal controller of water temperature. In combination with the outfit where the water tube can be embedded to allow warm water to go through, it can be used as a hot water treatment. It can provide a similar effect on the target portion of human body as a hot bath. This setup can be also used for a treatment for Peripheral Arterial Disease (PAD). A Purdue University Professor, Dr. Bruno Roseguini [45], clinically tested the effect of this therapy.

The key uniqueness of thermoelectric includes the potential compact packaging due to the flat structure of the module. This geometrical simplicity may provide an advantage compared to the off-the-shelf compact vapor compression cycle refrigerators, which are typically operated by AC power supply. The thermoelectric heat pump may enhance the portability in combination with a battery. Flat Li-ion batteries for smart phone may be applicable.

The heat pump mode operation of thermoelectric device is illustrated in Fig. 7.26. The electrical power W_{in} consumed to pump the heat Q_{th} from the colder side to the warmer side ends up with an additional joule heating $Q_{Joule} = W_{in}$ by the externally driven electrical current. The Joule heating occurs inside of the thermoelectric device and then this heat component is pumped to the warmer side. This results in the coefficient of performance (COP) on the heating side more than unity. In comparison, an electric ohmic heater performs COP equal to 1.0 or less:

$$\text{COP} = \frac{Q_{th} + W_{in}}{W_{in}} = \frac{SI\left(T_h - T_c\right) + I^2R}{I^2R} \qquad (7.16)$$

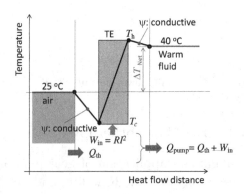

Fig. 7.26: Heat pump mode utilization of thermoelectric (TE) for medical/healthcare or clinical thermotherapy.

The interfaces on both sides of thermoelectric (TE) heat pump must have inverse temperate gradient as seen in the figure to allow the conductive heat flows ultimately from colder to warmer.

If the thermoelectric effect is small enough ($ZT \to 0$), the COP converges to unity where the heat pump effect becomes smaller and then the thermoelectric behaves equal to an electrical ohmic heater. If the electrical current I is very small, the COP is very large, but pumping capacity becomes very small. If I is infinitely large, COP $\to 0$. For more details on thermoelectric heat pump design, see Ref. [46].

7.5 Summary

Thermal energy recovery with small temperature gradient with thermoelectrics in wearable and IoT applications were discussed in detail. In these applications, a limited heat flux and only small temperature gradient are available. The design for maximizing the power output is essential for reasonably high power per cost. The fundamentals must follow the same model as power generation and cooling, while the uniqueness is in the impact of parameters, just as an example, thermal conductivity of the thermoelectric material and the one for gap filler. The materials are limited due to the requirement of mechanical flexibility. One approach is to design the TEG module with polymer-based material. Another is to design the structure flexible based on one-dimensional element. This section introduced these approaches and then analytically demonstrated the performance.

References

[1] S. Greenberg, E. Mills, B. Tschudi, P. Rumsey and B. Myatt, Best practices for data centers: Lessons learned from benchmarking 22 data centers, in *Proc. of the ACEEE Summer Study on Energy Efficiency in Buildings in Asilomar*, 76–87 (2006).

[2] S. P. Dharkar, K. Yazawa and E. Groll, Optimization of CO_2 heat pump system for simultaneous heating and cooling applications, in *Proc. of International Congress of Refrigeration*, #300 (2015).

[3] Birck nanotechnology center in discovery park at purdue university, Available at: http://www.purdue.edu/discoverypark/birck/.

[4] M. Ferrante and B. Brunone, Pipe system diagnosis and leak detection by unsteady-state tests, 1. Harmonic analysis, *Adv. Water Resour.*, **26**(1), 95–105 (2003).

[5] I. Stoianov, L. Nachman, S. Madden, T. Tokmouline and M. Csail, PIPENET: A wireless sensor network for pipeline monitoring, in *Proc. of the IEEE 6th Int. Symp. on In Information Processing in Sensor Networks*, 264–273 (2007).

[6] A. Faghri, *Heat Pipe Science and Technology*, Global Digital Press (1995).

[7] R. S. Prasher, A simplified conduction based modeling scheme for design sensitivity study of thermal solution utilizing heat pipe and vapor chamber technology, *Trans.-ASME J. Electronic Packaging*, **125**(3), 378–385 (2003).

[8] K. H. Do, S. J. Kim and S. V. Garimella, A mathematical model for analyzing the thermal characteristics of a flat micro heat pipe with a grooved wick, *Int. J. Heat and Mass Transfer*, **51**(19), 4637–4650 (2008).

[9] W. M. Keys and A. L. London, Chap. 6 Analytic solution for flow in tubes, in *Compact Heat Exchangers*, 3rd ed., Krieger Publishing Company, Florida, 113–139 (1984).

[10] T. Starner and J. A. Paradiso, Human generated power for mobile electronics, *Low Power Electron. Des.*, **45**, 1–35 (2004).

[11] H. L. Johnson, Wearable cooling respiratory device, US 3295522 A, (1967).

[12] E. W. Frantti, Thermoelectric air conditioning apparatus for a protective garment, US 3085405 A, (1963).

[13] G. B. Delkumburewatte and T. Dias, Wearable cooling system to manage heat in protective clothing, *J. Textile Institute*, **103**(5), 483–489 (2012).

[14] J. A. Paradiso and T. Starner, Energy scavenging for mobile and wireless electronics, *Pervasive Computing, IEEE*, **4**(1), 18–27 (2005).

[15] Beeby, S. P. and White, N. M., Low power systems (Section 5.4.1), In *Energy Harvesting for Autonomous Systems*, Artech House, 149–151 (2010).

[16] V. Leonov, T. Torfs, P. Fiorini and C. Van Hoof, Thermoelectric converters of human warmth for self-powered wireless sensor nodes, *Sens. J. IEEE*, **7**(5), 650–657 (2007).

[17] V. Leonov and R. J. M. Vullers, Wearable thermoelectric generators for body-powered devices, *J. Electron. Mater.*, **38**(7), 1491–1498 (2009).

[18] V. Leonov, T. Torfs, N. Kukhar, C. Van Hoof and R. Vullers. Small-size BiTe thermopiles and a thermoelectric generator for wearable sensor nodes, in *Proc. of the 6th Euro. Conf. on Thermoelectrics*, 10–12 (2007).

[19] V. Leonov, Thermoelectric energy harvesting of human body heat for wearable sensors, *IEEE Sens. J.*, **13**(6), 2284–2291 (2013).

[20] A. Harb, Energy harvesting: State-of-the-art, *Renew. Energy*, **36**(10), 2641–2654 (2011).

[21] A. P. Chandrakasan, D. C. Daly, J. Kwong and Y. K. Ramadass, Next generation micro-power systems, in *VLSI Circuits, 2008 IEEE Symposium on*, 2–5 (2008).

[22] R. JM. Vullers, R. Van Schaijk, H. J. Visser, J. Penders and C. Van Hoof, Energy harvesting for autonomous wireless sensor networks, in *IEEE Solid-State Circuits Magazine*, **2**(2), 29–38 (2010).

[23] L. Mateu, C. Codrea, N. Lucas, M. Pollak and P. Spies, Human body energy harvesting thermogenerator for sensing applications, in *IEEE Int. Conf. on Sensor Technologies and Applications*, 366–372 (2007).

[24] D. C. Hoang, Y. K. Tan, H. B. Chng and S. K. Panda, Thermal energy harvesting from human warmth for wireless body area network in medical healthcare system, in *Proc. of the 8th IEEE Int. Conf. on Power Electronics and Drive Systems* (2009).

[25] C. Watkins, B. Shen and R. Venkatasubramanian, Low-grade-heat energy harvesting using superlattice thermoelectrics for applications in implantable medical devices and sensors, in *Proc. of the 24th Int. Conf. on Thermoelectrics* (2005).

[26] S. E. Jo, M. K. Kim, M. S. Kim and Y. J. Kim. Flexible thermoelectric generator for human body heat energy harvesting, *Elec. Lett.*, **48**(16), 1015–1017, (2012).

[27] Kim, Sun Jin, Ju Hyung We and Byung Jin Cho., A wearable thermoelectric generator fabricated on a glass fabric, *Energy Environ. Sci.*, **7**(6), 1959–1965 (2014).

[28] I. Stark, Invited talk: Thermal energy harvesting with thermo life, in *IEEE International Workshop on Wearable and Implantable Body Sensor Networks*, 19–22, (2006).

[29] T. Park, C. Park, B. Kim, H. Shin and E. Kim, Flexible PEDOT electrodes with large thermoelectric power factors to generate electricity by the touch of fingertips, *Energy Environ. Sci.*, **6**(3), 788–792 (2013).

[30] O. Bubnova and X. Crispin, Towards polymer-based organic thermoelectric generators, *Energy Environ. Sci.*, **5**(11), 9345–9362 (2012).

[31] M. Chabinyc, Thermoelectric polymers: Behind organics' thermopower, *Nat. Mater.*, **13**(2), 119–121 (2014).

[32] M. Nakamura, A. Hoshi, M. Sakai and K. Kudo, Evaluation of thermopower of organic materials toward flexible thermoelectric power generators, *MRS Proc.*, **1197** (2009).

[33] R. J. M. Vullers, R. van Schaijk, I. Doms, C.Van Hoof and R. Mertens, Micropower energy harvesting, *Solid State Electron.*, **53**(7), 684–693 (2009).

[34] J. L. Monteith and L. E. Mount (Eds.), *Heat Loss from Animals and Man: Assessment and Control*, Elsevier (2013).

[35] J. H. Boyd and R. E. Mellman, The effect of fuel economy standards on the US automotive market: an hedonic demand analysis, *Transp. Res. A*, **14**(5), 367–378 (1980).

[36] J.-H. Bahk, H. Fang, K. Yazawa and A. Shakouri, Flexible thermoelectric materials and device optimization for wearable energy harvesting, *J. Mater. Chem. C*, **3**(40), 10362–10374 (2015).

[37] K. Yazawa and A. Shakouri, Scalable Cost/Performance Analysis for Thermoelectric Waste Heat Recovery Systems, *J. Elect. Mat.*, **41**(6), 1845–1850 (2012).

[38] Y. H. Kim, C. Sachse, M. L. Machala, C. May, L. Müller-Meskamp and K. Leo, Highly conductive PEDOT: PSS electrode with optimized solvent and thermal post-treatment for ITO-free organic solar cells, *Adv. Fun. Mat.*, **21**(6), 1076–1081 (2011).

[39] U. Lemmer S. Kettlitz, A. Gall and M. Gueltig, Wound and folded thermoelectric systems and method for producing same, US patent US9660167B2 (2012).

[40] OTEGO official web site, http://www.otego.de.

[41] G-H. Kim, L. Shao, K. Zhang and K. P. Pipe, Engineered doping of organic semiconductors for enhanced thermoelectric efficiency, *Nat. Mater.*, **12**, 719–723 (2013).

[42] M. Hokazono, H. Anno and N. Toshima, Thermoelectric properties and thermal stability of PEDOT:PSS Films on a polyimide substrate and application in flexible energy conversion devices, *J. Elect. Mater.*, **43**(6), 2196–2201 (2014).

[43] Snugovsky *et al.*, Low melting temperature solder alloy, US Patent Application 20130259738 (Mar 30, 2012).

[44] T. Park *et al.*, Flexible PEDOT electrodes with large thermoelectric power factors to generate electricity by the touch of fingertips, *Energy Environ. Sci.*, **6**(3), 788–792 (2013).

[45] D. Neff, A. M. Kuhlenhoelter, C. Lin, B. J. Wong, R. L. Motaganahalli and B. T. Roseguini, Thermotherapy reduces blood pressure and circulating endothelin-1 concentration and enhances leg blood flow in patients with symptomatic peripheral artery disease, *Am. J. Physiol. Regul. Integr. Comp. Physiol.*, **311**(2), R392 (2016).

[46] K. Yazawa and A. Shakouri, Optimum design and operation of thermoelectric heat pump with two temperatures, In *ASME 2015 International Technical Conference and Exhibition on Packaging and Integration of Electronic and Photonic Microsystems*, IPACK2015-48682, V001T09A080; 6, 2015.

Chapter 8

Materials and Device Characterization

In this chapter, several key techniques used for characterization of individual material properties as well as devices are reviewed. The measurement techniques for the three thermoelectric material properties, electrical conductivity, Seebeck coefficient, and thermal conductivity, of thin films and bulk materials are introduced and their limitations are discussed. For device characterization, various micro/nano-scale thermal characterization techniques are reviewed. In particular, thermoreflectance imaging, impedance spectroscopy, and the Harman method are discussed in detail for thermoelectric device testing.

8.1 Introduction

As thermoelectric devices gain popularity over the recent years, accurate characterization of the materials used and the devices in their final form has become more essential than ever for realization of high performance and reliable products. Material properties are highly dependent upon various parameters including synthesis conditions, doping levels, nano and microstructures in the material, as well as the temperature. Different measurement techniques are necessary based on the material dimensions. For instance, measurement of thin films requires special attention due to the existence of substrates and other layers affecting the measurement [1]. Low dimensional materials such as nanowires, quantum well films require special techniques suited for the specific material system. In the processes, a sophisticated calibration of various parameters needs to be re-done for each sample due to the variation in the parameters.

System performance varies substantially with surrounding conditions such as those of the hot and cold side interfaces, ambient air and gap fillers, which all in all constitute complex heat transfer within the system and cause parasitic heat losses to affect the performance. There are growing research activities in recent years that study the system performance variation under the actual operating conditions with various experimental techniques [2]. Three-dimensional finite element simulations are often combined with the experiments for accurate characterization along with the experimental efforts. Thermoreflectance imaging (TRI) is useful as a non-destructive, remote sensing technique for mapping of the surface temperature of a device under bias. This imaging technique can reveal thermoelectric performance at various locations on the device, as well as system reliability [3]. It is particularly useful for detecting thermal or electric faults and hotspots in the device by directly measuring the temperature rise in the device under operating conditions. Hotspots are critical for longevity of device as discussed in the previous chapters. Fast sensing is possible with the TRI technique to measure transient thermal responses with a sub 100 ns resolution. [4] Several other techniques used for micro/nano-scale thermal characterization such as micro-Raman thermography, and atomic force microscopy (AFM)-based techniques are also introduced and compared.

The Harman method is a well-known technique developed by T. C. Harman in 1958 [5] to study thermoelectric materials in a simple bar geometry, which allows simultaneous measurement of the figure of merit ZT and the constituting material properties. This technique takes advantage of the large difference between the thermal and electrical time constants of the system and requires precise adiabatic conditions at both sides of the system to be accurate. The air convection and parasitic heat losses through thermocouples must be minimized for temperature measurement. The Seebeck coefficient and the thermal conductivity cannot be extracted from this technique without prior knowledge of one of them.

Impedance spectroscopy is widely used in many scientific fields including battery science and so forth [6]. It has been recently utilized for thermoelectric characterization. The technique utilizes measuring the system impedance against AC inputs with varying frequency. As similarly shown in the Harman method, the difference between the thermal and electrical time constants causes separation of the thermoelectric properties in different frequency domains, and thus leads to extraction of the individual material properties without prior knowledge of any of them. We review the recently proposed methods based on impedance spectroscopy in the linear and the second-harmonic regimes for thermoelectric characterization.

8.2 Characterization of Individual Material Properties

8.2.1 *Seebeck coefficient*

As discussed throughout the book, the Seebeck coefficient, also called the thermopower, is a fundamental material property that relates the electron transport with the thermal transport in the material. It is defined as the ratio between the developed electric potential difference (ΔV) developed and the applied thermal potential difference (ΔT) in the same direction across the sample under no current flow ($I = 0$) as in

$$S = -\frac{\Delta V}{\Delta T} \tag{8.1}$$

Note that there is a minus sign added in Eq. (8.1) by the definition of Seebeck coefficient to make the Seebeck coefficient negative for n-type materials, and positive for p-type materials. The measurement of Seebeck coefficient involves the measurement of the two temperatures at the two different locations on the sample, and the measurement of voltage difference between the same two locations while a temperature difference is applied across the sample. Typically, temperature measurement is performed using thermocouples, as they are convenient to use and can be made small enough not to

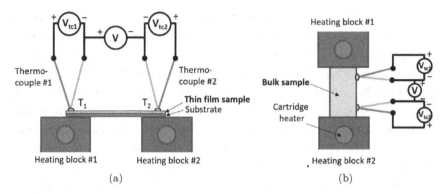

Fig. 8.1: Schematic of Seebeck measurement apparatus for (a) a thin film and (b) a bulk sample.

affect the temperature profile along the sample. Additional thermal paste can be used to ensure good thermal contact between the thermocouple junction and the sample surface, but it is important to minimize the use of the thermal paste. It is because the paste region can cause additional temperature drop and make the temperature measurement inaccurate. During the measurement of Seebeck coefficient, it is also important to keep an open circuit because the measured voltage will include the unwanted ohmic voltage when a current flows.

Figure 8.1 shows the schematics of typical Seebeck coefficient characterization systems for thin film and bulk samples. For thin film samples, temperature gradients are applied in the in-plane direction typically with two heating blocks sitting underneath the two sides of the sample for the in-plane Seebeck coefficient measurement [1]. Thermocouples are attached onto the thin film surface at the two locations for temperature measurements. These thermocouples can also be used as voltage probes as shown in Fig. 8.1. The voltage developed across the sample by the Seebeck effect is measured between the two thermocouple wires of the same polarity. However, this voltage includes the voltage developed along the thermocouple wires due to the difference between the temperatures at the thermocouple junction and at the other end of the wire. The corrected Seebeck coefficient can be found as

$$S = -\frac{V}{(T_1 - T_2)} + S_l \tag{8.2}$$

where S_l is the Seebeck coefficient of the thermocouple wire, which is known. The other two voltages (V_{tc1} and V_{tc2}) shown in Fig. 8.1(a) are converted to the temperatures (T_1 and T_2) by the thermocouple conversion chart or formula.

For the measurement of the cross-plane Seebeck coefficient of a thin film, a special attention is required because it is very difficult to make electric contacts at the two sides of the film across a small thickness. A substrate and other layers are typically involved between the two contacts, so that the measured voltage (and temperatures) include the contributions from these parasitic layers. A differential technique can be employed with two samples with and without the thin film, so that the parasitic contributions can be accounted for with the sample without the thin film. Also, due to the small thickness, the temperature difference in the cross-plane direction is very small. So is the Seebeck voltage. Therefore, the signals can be much smaller even than the noise levels. To overcome this difficulty, an AC method can be employed with lock-in frequency amplification to increase the signal-to-noise ratio. Details about this technique can be found elsewhere [1, 7].

For bulk samples, heating blocks are attached at the top and bottom sides of the sample as shown in Fig. 8.1(b) to ensure uniform one-dimensional temperature gradient along the length direction. Parasitic heat losses must be eliminated. Thermocouples are attached at the two positions at a side of the sample for temperature measurement, and the voltage is measured in the same manner as in the thin film case using two thermocouple wires of the same polarity. Typically, a thermally and electrically conducting interface material is used in between the heating block and the sample for good thermal and electrical contact. A pyrolytic graphite sheet is a good example of the interface contact material as it provides a high electrical conductivity ($>10^4$ S/cm) and high thermal

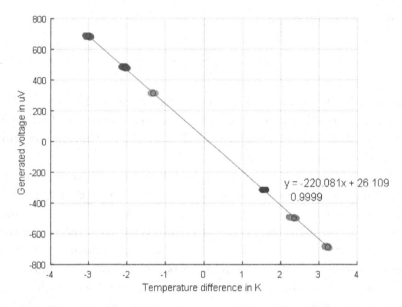

Fig. 8.2: Example plot of voltage vs temperature difference for the measurement of Seebeck coefficient of a Standard Reference Material® (SRM) No. 3451, Bi_2Te_{3+x}. The slope of the linear fit is the Seebeck coefficient $(= -220.08\,\mu V/K)$.

conductivity (>600 W/(m K)), which is higher than those of aluminum and cooper by several factors [8, 9].

Figure 8.2 displays an example plot of voltage as a function of temperature difference for the Seebeck coefficient measurement. The sample measured here is the standard reference material, SRM® 3451, which is a Bi_2Te_{3+x} of dimensions 3.5 mm × 2.5 mm × 8.0 mm. Six distinctive temperature differences were used for the measurement, and the slope of the linear fit of the data points is the Seebeck coefficient of the material. For each temperature difference, a sufficient time interval is provided before the measurement is taken until the temperature is stabilized, and multiple data points are collected for statistical analysis to account for the thermal noises.

8.2.2 *Electrical conductivity*

Measurement of electrical conductivity of a bulk sample with constant cross-sectional area is relatively simple as it simply requires a resistance measurement, and the measured resistance R can be converted to the electrical conductivity by

$$\sigma = \frac{1}{R}\frac{L}{A} \tag{8.3}$$

where L is the length of the sample, and A is the cross-sectional area of the sample. For accurate measurement, however, special care is needed in the measurement. Since Eq. (8.3) assumes a perfect 1D current flow in the length direction, a good electrical contact must be made at each side of the sample to cover the entire cross-section. A long bar-type sample with a relatively small cross-sectional area is desired for this reason. Also, the sample must be at a uniform temperature, i.e. no temperature gradient in any direction, to avoid the development of the Seebeck voltage. In addition, a small current must be used for the resistance measurement in order to minimize the Peltier effects at the two sides, which can create a temperature difference between the two sides, thus adding the unwanted Seebeck voltage.

8.2.2.1 *Van der Pauw method*

For thin film samples, in comparison to bulk samples, it is difficult to employ the same method because probes cannot be easily placed at the side wall of the film. Also, in most cases, the thin film sample can be in an arbitrary geometry, with, for example, a cross-section varying throughout the sample. In this case, the van der Pauw (vdP) method can be used for the thin film conductivity measurement. The van der Pauw method is a widely used four-probe method in which any arbitrary shape of thin film with a constant thickness can be measured for its electrical conductivity with four contacts created at the four corners on the top surface of the film. Figure 8.3(a) shows a typical van der Pauw measurement example with four contacts. [10, 11] There is no requirement about the sample geometry for van der Pauw measurement, except that the four contacts must be much smaller than the sample size. Yet, there is a correction factor that can be used for finite sizes of contacts for specific sample geometries discussed in the original paper [11].

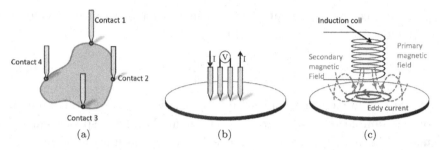

Fig. 8.3:   Schematics of (a) an arbitrary van der Pauw (vdP) geometry, (b) the collinear four-probe method, and (c) the non-contact eddy current technique for electrical conductivity measurement of thin films.

The van der Pauw conductivity measurement involves two resistance measurements: one measurement is taken with an electric current injected between the two adjacent contacts, say Contacts 1 and 2, and the measurement of voltage between the two other contacts, Contacts 3 and 4, shown in Fig. 8.3(a), yielding the first resistance $R_{34,12} = \frac{V_{34}}{I_{12}}$. The second measurement is taken with another electric current injected between another two adjacent contacts, i.e. Contacts 2 to 3, with voltage measurement between Contacts 4 and 1 to get $R_{41,23} = \frac{V_{41}}{I_{23}}$. Using the two measured resistances in the vdP equation, Eq. (8.4), one can obtain the film electrical conductivity σ.

$$\exp\left(-\pi R_{34,12}\sigma d\right) + \exp\left(-\pi R_{41,23}\sigma d\right) = 1 \qquad (8.4)$$

where d is the film thickness. Equation (8.4) needs to be solved numerically to determine σ.

In practice, each resistance measurement in the vdP method can be affected by the Seebeck voltage induced by the temperature variation on the sample. In order to eliminate the unwanted Seebeck voltage, one can flip the current direction and measure the resistance twice with the two current directions, and average them out to obtain each resistance, $R_{34,12}$ or $R_{41,23}$. With the flipped current direction, the direction of the ohmic voltage will be flipped while the Seebeck voltage remains the same. By averaging the two resistances with two opposite current directions, the Seebeck voltage can be removed from the ohmic one.

8.2.2.2 *Collinear four-probe method*

Other than the vdP method, the collinear four-probe method is also widely used for thin film electrical conductivity measurement [12]. In this method, four probes with equal spacing are put in contact with the sample surface as schematically shown in Fig. 8.3(b), and a current is injected between the two outer probes and the voltage in between the two inner probes is measured. The in-plane electrical conductivity of the sample is then obtained by

$$\sigma = \frac{\ln 2}{\pi d} \frac{I}{V} \tag{8.5}$$

Note that Eq. (8.5) is valid only if the sample size is much larger than the probe spacing, and the thickness is much smaller than the probe spacing. Otherwise, correction factors can be employed to account for the non-uniform current flow [13, 14].

8.2.2.3 *Non-contact eddy current method*

The aforementioned four-probe techniques require the probes to be in direct ohmic contact with the sample surface. The contact of the probes often needs elaborate sample processing with surface metallization and photolithography. Also, it causes damage to the surface due to the physical contact with sharp probe tips. Non-contact techniques are, therefore, useful to avoid such complications in sample preparation, and sample damages. Many kinds of non-contact techniques have been developed thus far. As one of them, the interactions of the sample material with electromagnetic fields from radio frequency (RF) to microwave frequencies are measured and analyzed to obtain the conductivity of the sample [15]. The capacitive techniques are widely used for conductivity measurement of highly resistive non-magnetic materials, such as lowly doped semiconductors, in both frequency domain and time domain. These techniques measure the effective capacitance of the sample, and the R-C product from the frequency domain cut-off frequency or from the time constant of the transient voltage discharging in the time domain, both of which together gives the resistance of the sample, and thus the electrical conductivity [15, 16].

For non-magnetic, highly conducting materials such as metals and degenerate semiconductors, the RF eddy current technique is commonly used as a standard non-contact resistivity measurement. This technique has been used for resistivity measurement of metals for a long time, and was implemented for semiconductors in 1976 by Miller *et al.* [17] at Bell labs. A schematic of a typical RF eddy current measurement system is shown in Fig. 8.3(c). The induction coils is placed in proximity of, but not in direct contact with the sample, so that it induces magnetic fields in the sample. This magnetic field generates eddy currents in the sample that dissipate the electromagnetic energy supplied by the RF current generator. Assuming no flux leakage and negligible surface effects, the power dissipated in this process is proportional to the sheet conductance of the sample. The power loss can be measured with an eddy current sensor in which the reduction in the magnetic field due to the opposing secondary field created by the eddy current is detected by the voltage drop in the coil sensor. This power is finally converted to the sheet conductance of the sample after careful calibration of the correlation.

8.2.3   *Thermal conductivity*

The thermal conductivity is a fundamental material property that represents the ability of the material to conduct heat. By the Fourier law, it is defined as the proportionality factor of the temperature gradient in relation to the heat flux as in

$$\vec{q} = -\kappa \nabla T \tag{8.6}$$

In principle, the thermal conductivity is a function of position in a non-homogenous material, but the 'average' value of thermal conductivity of a bulk material across the cross-sectional area A and length L can be obtained by measuring the total heat input Q, and the temperature difference applied across the length of the sample simply as

$$\kappa = \frac{Q}{\Delta T} \frac{L}{A} \tag{8.7}$$

This measurement technique, however, can impose a large uncertainty because both heat input and temperature difference are difficult to measure accurately. Since the heat can easily spread everywhere every direction, i.e. there is no perfect thermal insulator, it is extremely difficult to accurately quantify the heat input. Parasitic heat losses such as those from the convective and radiative heat transfer from the sample surface to the surroundings must be taken into account to determine the heat input. Most of the measurement is thus conducted in high vacuum to minimize the convection heat losses, and often with radiation shields to minimize the radiation heat losses as well, which can be significant at high temperatures. Thermocouple temperature measurement is also non-trivial because once the thermocouples are attached, it can alter the temperature due to the parasitic heat conduction through the thermocouple wires. It is therefore desirable to use thermocouples with wire diameters much smaller than the sample size, and with low thermal conductivity wires, e.g. chromel and constantan of type-E thermocouples.

In order to overcome these limitations, many advanced techniques have been developed for thermal conductivity measurement. Yet, each technique has its own limitations, and thus one must select an appropriate technique suited for the specific material and samples to measure. Selection criteria are based on many factors including sample geometry and dimensions, related thermophysical properties of the material and their ranges, the requirement of sample preparation, and potential error sources related to parasitic heat losses and temperature non-uniformity.

Comprehensive review of the available measurement techniques is beyond the scope of this chapter. Herein, we briefly discuss a few selected commonly used techniques: the comparative steady-state method for bulk materials, the laser flash method, 3-ω method, and time-domain thermoreflectance (TDTR). There are excellent recent reviews elsewhere with more information about these techniques and others [18, 19].

8.2.3.1 *Comparative steady-state method*

In the comparative steady-state method, the sample to be measured and a reference sample with known thermal conductivity are

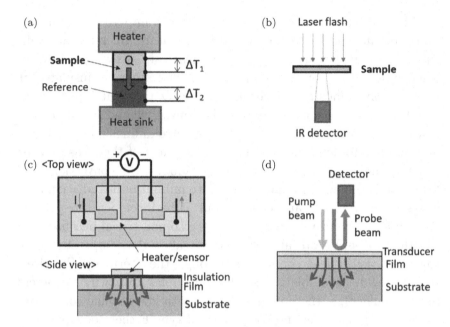

Fig. 8.4: Schematics of key thermal conductivity measurement techniques: (a) comparative steady-state bulk measurement, (b) the laser flash method, (c) 3-ω method, and (d) time domain thermoreflectance (TDTR) technique.

compared in terms of temperature difference across them for a constant heat flow to obtain the thermal conductivity of the sample. There is no need of heat flow measurement, which is one of the major error sources otherwise. The sample and reference are placed together in series in the heat flow direction between a heater and a heat sink, as schematically shown in Fig. 8.4(a). Assuming there is no heat leakage, the same heat must flow through the sample and the reference. From the 1D Fourier law, the thermal resistance is then directly proportional to the temperature difference applied in each sample. By measuring the two temperature differences, i.e. ΔT_1 and ΔT_2 in Fig. 8.4(a), one can obtain the thermal conductivity of the sample as

$$\kappa = \frac{\Delta T_2}{\Delta T_1} \frac{L_1}{L_2} \frac{A_2}{A_1} \kappa_2 \tag{8.8}$$

where L and A are, respectively, the length and cross-sectional area of the sample in the flow direction, and the subscripts 1 and 2 indicate the properties of the real sample, and the reference, respectively. This technique is simpler, but, as discussed above, heat leakage due to convection and radiation from the side wall of the sample, as well as due to the conduction through the thermocouple wires used for temperature measurement must be minimized to be accurate.

8.2.3.2 *Laser flash method*

In order to overcome these limitations and large uncertainties in the steady-state techniques, many techniques based on transient thermal responses have been developed. Pulses and sinusoidal heat inputs are used in the transient techniques, and the fast measurements of transient temperature changes are monitored to obtain the thermal characteristics of the sample. One of the most widely used such techniques is the laser flash method. The laser flash technique is a non-contact, non-destructive method to measure the thermal diffusivity [20]. To obtain the thermal conductivity, one must know or determine the mass density and the specific heat of the material by separate experiments, such that

$$\kappa = \alpha \rho C_p \qquad (8.9)$$

where α is the thermal diffusivity, ρ is the mass density, and C_p is the specific heat of the material. The specific heat can be measured using differential scanning calorimetry (DSC) [21].

A schematic figure of the laser flash method is displayed in Fig. 8.4(b). This technique utilizes remote optical heating with a laser flash and measures the transient temperature response using a remote infrared (IR) detector from the backside of the sample. No sample preparation is required except for the sample cutting into desired dimensions due to the non-contact nature of the technique. Normally, a disk sample is used. Due to the thermal diffusion, it takes time for the heating at the top surface of the sample to reach the backside. Assuming a thermally insulated sample, the time that it takes for the backside temperature to rise to one half of the maximum (steady-state) temperature is inversely proportional to the thermal diffusivity of the sample, and quadratically proportional to the thickness. With a multiplication constant, the thermal diffusivity is given by [20]

$$\alpha = \frac{1.38}{\pi^2} \frac{d^2}{t_{\text{half}}} \tag{8.10}$$

where t_{half} is the time that takes for the backside to reach one half of the maximum temperature, and d is the thickness of the sample.

Since the density, specific heat, and diffusivity must be measured separately, the errors from these individual measurements get multiplied to potentially create a large uncertainty in the final thermal conductivity value [19]. Also, the heat leakage and influence by the sample holder can cause an error in the diffusivity measurement. The thickness requirement for the laser flash method is dependent upon the time scales associated with the laser pulses and the infrared detector. Typically, a thickness larger than 100 μm can be measured by this method, but the sample thickness can be thinner for a material with a small thermal diffusivity. For much thinner films below 100 μm, the 3-ω method can be used instead, which is discussed in the following section.

8.2.3.3   3-ω method

The 3-ω method has been widely used for thermal conductivity measurement of thin films and bulks since it was introduced by D. G. Cahill in 1990 [22]. Figure 8.4(c) shows a typical sample structure with a metal heater/sensor strip. The metal strip is microfabricated on top of the sample, and four contact pads are created along the heater line as shown in the figure for current injection and voltage measurement. Typical width of the metal strip is 10–100 μm, and the length must be much larger than the width for the assumption of an infinitely long device. This metal strip is used as a heater and a temperature sensor at the same time. A sinusoidal current $I = I_0 \cos(2\omega t)$ is injected into the heater strip, which creates Joule heating along the strip. This Joule heating induces the temperature rise in the heater at the second harmonic, i.e. 2ω, because it is proportional to the Joule power, $I^2 R$, where I^2 term creates the second harmonic and the change in R is much smaller, thus ignored, so that

$$\Delta T(t) = \Delta T_0 \cos\left(2\omega t + \varphi\right) \qquad (8.11)$$

where ΔT_0 is the amplitude of temperature change, and φ is the phase delay. The temperature rise ΔT_0 can be measured by measuring the slight change in resistance of the strip, as the resistance of a metal increases linearly with temperature for a small temperature change. The relation between the rise of resistance and temperature must be calibrated for the metal strip beforehand, and is represented by the parameter called the temperature coefficient of resistance (TCR), which is defined as $\beta_{TCR} = \frac{1}{R}\frac{dR}{dT}$. Then, the resistance rise due to the Joule heating is obtained as

$$R(t) = R_0(1 + \beta_{TCR}\Delta T) = R_0(1 + \beta_{TCR}\Delta T_0 \cos\left(2\omega t + \varphi\right)) \qquad (8.12)$$

One cannot directly measure the resistance change to obtain the temperature rise. One does so by measuring the voltage instead between the other pairs of contacts. The voltage is, by the Ohmic law, IR, such that

$$V(t) = I(t)R(t) = I_0 R_0 \cos(2\omega t) + \frac{1}{2}I_0 R_0 \beta_{TCR}\Delta T_0 \cos(\omega t + \varphi)$$

$$+ \frac{1}{2}I_0 R_0 \beta_{TCR}\Delta T_0 \cos(3\omega t + \varphi) \qquad (8.13)$$

According to Eq. (8.13), the temperature rise ΔT_0 appears in both 1ω and 3ω terms at the same time. However, in the 1ω term, another term, which is the first term on the far right-side of Eq. (8.13), also appears, and is much greater than the second term that contains ΔT_0. Therefore, it is not feasible to directly measure ΔT_0 from the 1ω voltage. Instead, by measuring the amplitude of voltage at 3ω, which is directly proportional to ΔT_0, ΔT_0 can be obtained simply by

$$\Delta T_0 = \frac{2}{I_0 R_0 \beta_{TCR}} V_{3\omega} \qquad (8.14)$$

Yet, the 3ω term is much smaller than the first harmonic voltage by several orders of magnitude. Thus, a lock-in amplifier is used to precisely measure the 3ω signal.

Note that this temperature rise includes both those in the thin film and the substrate. One can do the same measurement with another sample having everything the same as in the first sample, but only without the thin film. From this second measurement, one can obtain the temperature rise due to the substrate only, i.e. ΔT_{sub}, and thus the temperature difference across the thin film only becomes $\Delta T_{\text{film}} = \Delta T_0 - \Delta T_{\text{sub}}$.

If the heater width is much larger than the film thickness, the edge effects will be negligibly small, and therefore, the heat flow in the film can be assumed to be one-dimensional, i.e. uniformly in the vertical direction. In this case, the 1D Fourier law holds, so that $\Delta T_{\text{film}} = Q\left(\frac{d}{\kappa A}\right)$, where Q is the heat flow being equal to the Joule power $\left(=\frac{1}{2}I_0^2 R_0\right)$, and d is the thickness of the film, and $A(=Lw)$ is the area of the heater strip. Finally, the thermal conductivity is obtained by

$$\kappa = \frac{I_0^2 R_0 d}{2A\Delta T_{\text{film}}} \tag{8.15}$$

An additional note is that the thermal conductivity of the substrate cannot be determined in the same manner with ΔT_{sub}, because the heat flow in the substrate is highly two-dimensional due to the spreading, rather than 1D. Instead, one can do frequency-varying measurements and extract the thermal conductivity of the substrate. Within the line heater approximation where the width of the heater strip is much smaller than the dimensions of the substrate, the temperature rise in the substrate, ΔT_{sub}, decreases linearly with increasing ω in the logarithmic scale, or $\ln \omega$, in the frequency range where the heat penetration depth is smaller than the substrate thickness, typically within $100\,\text{Hz} \sim$ a few kHz. This is because the heat penetration depth decreases with increasing ω, or in other words, the oscillation of heating power turns heat back down more quickly before the temperature increases higher at a higher frequency. In this frequency range, the thermal conductivity of the substrate can be additionally determined by the linear slope of ΔT_{sub} with respect to $\ln \omega$ such that [23]

$$\kappa_{\text{sub}} = \frac{I_0^2 R_0}{2\pi L} \left[-\frac{\partial \Delta T_{\text{sub}}}{\partial \ln(\omega)} \right] \qquad (8.16)$$

where L is the length of the heater strip between the two voltage probes.

8.2.3.4 *Time domain thermoreflectance*

Lastly, time domain thermoreflectance (TDTR) is another widely used, non-contact, non-destructive, and transient technique to measure thermal conductivity of films and bulks [24]. Normally, a thin metal layer is deposited on top of the sample as a transducer. This transducer absorbs the optical pulses called the 'pump' signals, and converts the energy to heat, which then spreads through the film and the substrate underneath the transducer layer. The transient heating is detected by another optical signal called the 'probe' signal, of which the reflectivity off the surface changes with the surface temperature. By careful calibration, the change in reflectivity can be converted to the temperature rise.

The pump pulses are normally extremely narrow, coming from a femtosecond laser, so that it can be assumed to be a delta function-like heating. By precisely controlling the time delay between the pump and probe signals, one can capture the transient thermal response as a function of time for the impulse heating. The pump signal is also modulated at modulation frequency ω_0, and the thermoreflectance response is measured for both the real part (in-phase) and the imaginary part (out-of-phase) as a function of time delay. The in-phase signal V_{in} represents the change in surface temperature. The peak in the V_{in} represents the surface temperature rising right after the pump pulse arrives, and the decaying tail in V_{in} represents the cooling of surface due to the heat dissipation in the sample. The out-of-phase signal can be viewed as the sinusoidal heating of the sample due to modulation at ω_0. Typically, the ratio between the in-phase and out-of-phase signals, i.e. $V_{\text{in}}/V_{\text{out}}$, as a function of time is fitted using a heat transfer model for varying modulation frequencies. The fitting parameters determine the thermal conductivity of the film, and the interface thermal resistance between the film and the substrate. More details about TDTR measurement and analysis can be found in many excellent papers [24–26].

8.3   Micro/Nano-scale Thermal Characterization Techniques

Various techniques for thermal characterization of thin films at the micron and submicron length scales have been developed over past few decades. Several key techniques are summarized in Table 8.1 and discussed in the following subsections.

8.3.1   *Micro-Raman thermography*

Micro-Raman thermography is a non-contact method based on the analysis of Raman scattered laser light in semiconductors or insulators. Sample surface is scanned by a focused laser beam and the ratio of Stokes and Anti-stokes peaks of the scattered light are analyzed to obtain a surface temperature map. Temperature resolution is based on the experimental calibration of the Raman spectra, and the accuracy has been reported in the range of a few K [27]. Spatial resolution below 1 μm has been reported [28]. A 10 ns temporal resolution of a time-resolved micro-Raman has been reported as well [29]. Micro-Raman excels at measuring semiconductors with some capacity to measure temperature below the surface, but is typically unable to measure metals. Recently, microparticles that have strong temperature-sensitive Raman signals such as TiO_2 have been deposited on the device to measure the surface temperature made of metals [30].

8.3.2   *Atomic force microscopy-based thermography*

Various atomic force microscopy (AFM)-based techniques have been adapted to measure temperature with very small spatial resolutions of 1 to 10 nm. Among them, scanning thermal microscopy (SThM)

Table 8.1: Comparison of various micro/nanoscale thermal characterization techniques.

Measurement technique	Contact/ Non-contact	Thermal resolution	Spatial resolution	Temporal resolution	Scan speed	Limitations	Ref.
Micro-Raman thermography	Non-contact	~1 K	<1 μm	<10 ns	Slow	Unable to measure metals, low T resolution	[27–29]
Scanning thermal microscopy (SThM)	Contact	N/A	<100 nm	N/A	Slow	Slow scanning, contact-related artifacts	[31–33]
Scanning Joule expansion microscopy (SJEM)	Contact	N/A	<10 nm	N/A	Slow	Slow scanning, contact-related artifacts	[35, 36]
Near-field scanning optical microscopy (NSOM)	Near contact	N/A	<20 nm	N/A	Slow	Slow scanning, contact-related artifacts	[37, 38]
Infrared thermal imaging	Non-contact	<50 mK	<5 μm	~20 μs	Fast	Spatial resolution limited by optical diffraction	
Thermoreflectance imaging (TRI)	Noncontact	~10 mK	<200 nm	<100 ns	Fast	Calibration of C_{TR} required	

and scanning Joule expansion microscopy (SJEM) have been popular. SThM measures the temperature-induced potential difference (thermovoltage or Seebeck voltage) between the AFM cantilever probe tip and the sample surface, or within the tip itself with a two-wire thermocouple [31, 32]. A variation of SThM measures electrical resistance instead of thermovoltage [33]. Spatial resolutions of SThM and other AFM-based techniques are determined by the size of the probe tip, and are typically ~100 nm or smaller. Improvement in thermocouple tip fabrications has enabled SThM temperature measurement with spatial resolution approaching ~10 nm [34]. SJEM is similar to SThM but achieves superior spatial resolution without the need to fabricate an extremely small probe tip. This technique scans a conventional AFM probe tip across the sample surface and extracts the temperature distribution directly from the Joule expansion of the sample in response to a lock-in excitation signal. [35] Spatial resolution is 1 nm to 10 nm, similar to other AFM-based techniques. Temperature resolution of 250 mK has been reported [36]. Despite the ultra-high spatial resolution, the AFM-based thermometry techniques suffer from contact-related artifacts such as heat flux variation due to poor contact. Also, owing to the nature of AFM technique, the scanning process is very slow, so the techniques are practical only for measuring a very small surface area of the sample.

8.3.3   *Near-field scanning optical microscopy*

Near field scanning optical microscopes (NSOM) use an optical fiber or a nano-antenna instead of a probe tip to induce temperature-dependent near-field light scatterings [37]. By bringing the fiber or antenna tip within 1 nm to 10 nm of the sample, spatial resolution is no longer constrained by the far field diffraction limit, and is determined by the aperture size on the metal coated fiber tip. A 12 nm resolution has been reported [38]. NSOM can measure temperature using optical signal reflectance, Raman spectroscopy, or infrared thermometry. As in other AFM-based techniques, it is crucial to understand the heat transfer properties between tip and sample in NSOM. This heat transfer controls spatial and temperature resolution, and measurement artifacts [39].

8.3.4 *Infrared thermal imaging*

Infrared thermal imaging is one of the popular methods for thermal characterization of microstructures [40]. The method measures temperature-dependent radiation from semiconductors and metals. An emissivity scaling factor is used to compensate for each material's non-ideal surface deviating from blackbody. Commercial imaging systems are now common, which capture microscopic thermal images quickly in a relatively simple measurement procedure. The method is all optical, remote, non-contact, and does not require special sample preparation. Spatial resolution in far-field optical systems is determined by the Rayleigh diffraction limit, $d = \lambda/(2NA)$, where d is the smallest resolvable feature, λ is the infrared wavelength, and NA is the numerical aperture of the lens. Most infrared imaging microscopes operate in the 3–5 μm range of wavelength and produce spatial resolution on the same scale. Systems working in near infrared (NIR) wavelengths, 0.8–3 μm, can achieve spatial resolution of $\sim$1 μm [41, 42]. However, near infrared systems are limited to measuring radiation corresponding to temperatures above 500 K, as the signal at these shorter wavelengths becomes too small at lower temperatures. Temperature resolution for commercial systems is typically in the range of 50 mK, but 10 μK has been achieved using a lock-in infrared thermal microscope [43]. Transient infrared microscopy imaging has been demonstrated with 20 μs time resolution using internal camera shuttering [44].

8.3.5 *Thermoreflectance imaging*

Thermoreflectance imaging (TRI) is a non-invasive optical technique for surface temperature mapping of electronic and thermoelectric micro-devices and thin film materials [3]. Figure 8.5 shows the principle of a TRI system [4]. The LED flash is illuminated onto the sample surface through an optical microscope, which is reflected back from the surface, and then captured by a high performance charge-coupled device (CCD) camera equipped in the microscope. The reflectivity of the LED light is measured by the CCD camera and converted to a relative temperature rise from background using the thermoreflectance

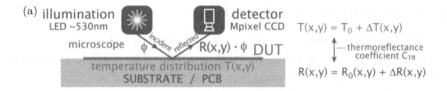

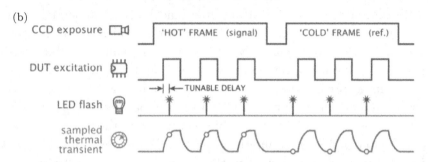

Fig. 8.5: (a) Principles of CCD camera-based thermoreflectance imaging and (b) schematic timing diagram for ultra-fast transient thermal characterization.

Source: Reproduced with permission from *J. Appl. Phys.*, **113**, 104502 (2013).

coefficient C_{TR} defined as the relative change in reflectivity with respect to a temperature change:

$$C_{\text{TR}} = \frac{1}{R_{\text{LED}}} \frac{\partial R_{\text{LED}}}{\partial T} \tag{8.17}$$

The thermoreflectance coefficient is a basic material property that depends on illumination wavelength, ambient temperature, material surface characteristics such as surface roughness. Therefore, C_{TR} must be calibrated for the specific sample surface before the actual measurement. For most metals and semiconductors of interest, the value of C_{TR} is in the order of 10^{-2} K^{-1} to 10^{-5} K^{-1} [45].

Since the entire 2D camera image can be directly converted to a surface temperature map, the 2D temperature distribution across the sample surface can be quickly obtained. Multiple identical images are collected through a train of LED pulses to average out the background noises. With a precise control of the delay between the onset time of device excitation (pump) and the LED flash time (probe) as shown in Fig. 8.5(b), ultra-fast transient thermal imaging is possible.

This method is sensitive to mK temperature fluctuations with sub-microsecond temporal resolution, and can produce images with sub-micron spatial resolution [46, 47].

TRI technique is very effective in diagnosing device behavior. Current leaks and material non-uniformity are easily identified by capturing hotspots in the device through TRI. Gate failure in MOSFET devices, and current leakage in integrated circuits have been detected by TRI [48,49]. Hotspots with high heat flux local heating in graphene and silver nanowire networks have been observed at the bottleneck of percolation [50]. The localized self-heating reroutes the percolation path and even irreversibly damages the material. Therefore, the TRI experiments are crucial for the study of robust electrodes based on nanowire/nanotube networks. Transient thermal response by TRI can additionally reveal the important physics of thermal transport along the electronic devices and materials, particularly being able to capture the evolution of hotspots and heat spreading that is happening in sub-microsecond time scales.

Furthermore, TRI technique can be used to separate the Peltier heating/cooling from the Joule heating in the micro-material/device using sinusoidal device excitations. In this method, a sinusoidal current is supplied to the device under test. Peltier and Joule terms will be observed at the first and second harmonic of the excitation frequency, respectively [4]. Figure 8.6(a) displays an example of separated Peltier and Joule signals acquired by this technique for a microscale thermoelectric cooler [4]. Separated Peltier signals can be used to reveal the localized thermoelectric properties in the material/device, i.e. the Seebeck coefficient mapping over the surface as well as the interface properties between grains in nanostructured materials.

Although TRI is a surface thermography technique, thermal characterization of a deep channel under the surface may be possible if the surface thermal image is reinforced with additional 3D finite-element thermal simulations. Transient self-heating and heat dissipation in a 0.2 μm wide active channel of a gate-all-around transistor covered by a gate oxide and multiple gate metal layers were characterized by this method, as shown in Fig. 8.6(b) [51].

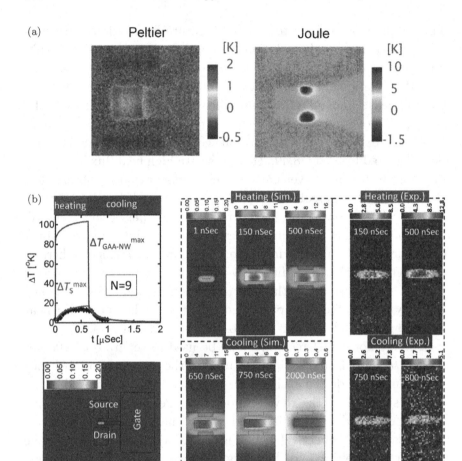

Fig. 8.6:　(a) Separated Peltier and Joule signals on a thermoelectric micro-cooler by thermoreflectance imaging technique and (b) finite-element thermal simulation and thermoreflectance imaging of a III-V gate-all-around transistor to estimate the temperature inside the channel under the gate layer during operation.

Source: (a) Reproduced with permission from *J. Appl. Phys.*, **113**, 104502 (2013); (b): Reproduced with permission from *IEEE Trans. Elect. Dev.*, **62**, 3595 (2015).

8.4 Harman Method

The Harman method was proposed by T.C. Harman in 1958 [5], in which a simple homogeneous bar-type sample shown in Fig. 8.7(a) is applied with an electric current in adiabatic boundary conditions for direct measurement of the thermoelectric figure of merit of the material. Due to the current flow, the Peltier effects at both sides create a temperature gradient along the sample. At steady state with a DC current, the heat transported as a result of the Peltier effects is equal to that transported in the opposite direction by thermal conduction that is caused by the temperature gradient. Assume that the Joule heating is much smaller than these two components of heat, such that

$$STI = \kappa \Delta T \left(\frac{A}{L} \right) \tag{8.18}$$

where S is the Seebeck coefficient, T is the absolute temperature, I is the current applied, κ is the thermal conductivity, ΔT is the temperature difference created between the two sides of the sample by the Peltier effect, A is the cross-sectional area, and L is the length of the sample. In this DC current measurement, the DC voltage across

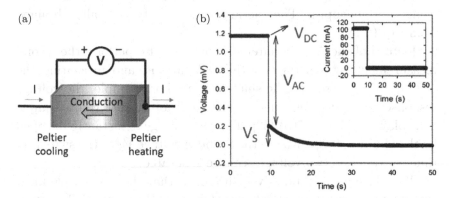

Fig. 8.7: (a) Schematic of the Harman method for direct ZT measurement of a p-type sample. For n-type, the temperature gradient and the heat flow will be in the opposite direction. (b) Voltage response with a DC current switch-off for an n-type skutterudite sample.

Source: (b) Reproduced with permission from *J. Appl. Phys.*, **125**, 025111 (2019).

the sample includes the ohmic voltage and the Seebeck voltage created by the temperature gradient, i.e. $V_{DC} = V_{Ohmic} + V_S$.

Another measurement is performed to separate the two voltages in which an AC current of the same amplitude is applied instead of DC, i.e. 60 Hz as in the original paper. For such an alternating current, the Peltier heating and cooling are switched between the two sides of the sample so quickly that the temperature gradient cannot be created in the sample. Then the voltage drop in the sample is solely ohmic, $V_{AC} = V_{Ohmic} = IR$. Thus, the Seebeck voltage is obtained by subtracting the AC voltage from the DC one:

$$V_S = S\Delta T = V_{DC} - V_{AC} \tag{8.19}$$

From these two DC and AC measurements, the figure of merit ZT can simply be obtained by

$$ZT = \frac{S^2}{\rho\kappa}T = \frac{V_S}{V_{AC}} = \frac{V_{DC} - V_{AC}}{V_{AC}} \tag{8.20}$$

Separate determination of S, σ, and κ are also possible from this method. The electrical conductivity σ can be obtained from the ohmic (AC) voltage, $V_{AC} = IR = I\frac{L}{\sigma A}$ as the sample dimensions and the current amplitude are known. The Seebeck coefficient S is obtained from Eq. (8.19) by measuring the temperature difference using thermocouples. The thermal conductivity is finally obtained from Eq. (8.18).

Instead of the AC measurement, one can also perform the second measurement by shutting off the DC current, and measuring the instant voltage drop at the shut-off time. For example, the voltage response during the DC current switch-off for a skutterudite sample is displayed in Fig. 8.7(b) [52]. The DC voltage is instantly dropped by the ohmic voltage, V_{AC} to the Seebeck voltage V_S at the switch-off time, so that the two voltages can be separated.

The Harman method is very useful in that the two simple measurements can determine the figure of merit of the material. To obtain accurate results, however, several considerations must be taken into account. First, the analysis assumes adiabatic conditions around the sample. But there can be significant conduction heat losses through the thermocouple wires and electrical leads attached for temperature and voltage measurement. Also, there can be heat losses through convection at the side wall of the sample. Therefore, smaller-diameter (larger gauge) thermocouples and leads are preferred to minimize the conduction losses, and the measurement needs to be performed in high vacuum to minimize ambient air convection. Also, the sample must be hanging in the vacuum chamber in no contact with any sample stages to achieve adiabatic conditions on all sides. It may be necessary to hold and support the sample in vacuum with rigid electric probes used for voltage measurement.

Another important source of error could be from the assumption that the Joule heating is negligibly small in the sample. For relatively resistive materials, this assumption may not hold. Kobayashi *et al.* [53] proposed a modified Harman method to overcome this Joule heating issue. In their method, square pulse trains are used for AC measurement instead of sinusoidal currents, as shown in Fig. 8.8(a). By switching between $+I$ and $-I$ with the same magnitude and duration, the Joule heating remains constant in the sample as in the DC case with continuous $+I$ current because Joule heating is independent of the current direction (Fig. 8.8(b)). If a sinusoidal current oscillation is used, the Joule heating will also be oscillated in the sample, which makes the analysis difficult. Under adiabatic condition, the uniform Joule heating by square current oscillation will increase the temperature uniformly across the sample, so that it does not create any temperature gradient. So, the voltage rise with DC current can be attributed to the Peltier effect only, hence, $V_S = S\Delta T$ still holds. Yet, the uniform temperature rise during the square AC measurement induces a voltage rise, as shown in Fig. 8.8(c), because the resistance increases with temperature, so that $V_{AC} = IR$ increases. However, the increase in R is usually relatively small compared to the absolute magnitude of R when a sufficiently small current is used in a few mA. So, the measurement of V_{AC} can be still accurate. This AC voltage increase by the Joule heating, however, could have affected the Seebeck voltage, because the latter is relatively small compared to the ohmic voltage. Therefore, $V_S = V_{DC} - V_{AC}$ can be accurately measured using this modified Harman method.

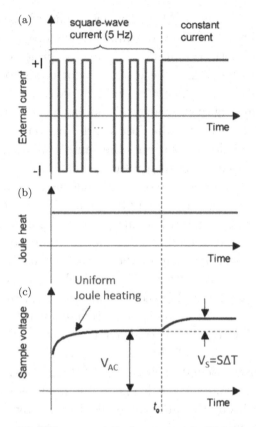

Fig. 8.8:   The modified Harman method proposed by Kobayashi *et al.* [53]. (a) Alternating square current (AC) and DC current inputs used, (b) induced constant Joule heating in the sample due to the AC/DC currents, and (c) voltage evolution with time corresponding to the current input.

8.5   Impedance Spectroscopy

Recently, there have been significant efforts to utilize the frequency-dependent impedance responses with AC current inputs to characterize TE systems [51–53]. This method is similar to the Harman method described earlier, but generates more information regarding the impedance response in the complex coordinates over a broad range of frequency. Dihlaire *et al.* [54] showed that all the thermoelectric properties of a TE material or device (S, σ, κ, and ZT) can be extracted from one single impedance spectroscopy measurement provided that the heat capacity is known. An analytical model based on the thermal quadrupole method was used to fit the impedance spectrum to extract the property values in this work. The impedance spectra have also been fitted using the equivalent lumped electrical circuit models [51]. Since the impedance spectrum of a TE system depends not only on the material properties of the system, but also on the contacts, heat loss by convection, and the boundary conditions (adiabatic or constant temperature), this technique can be very useful for characterizing both material properties and device parameters such as contact resistances in different operating conditions. For accurate material characterization, however, it is essential to perform the impedance spectroscopy in a high vacuum to minimize the parasitic convection heat losses.

García-Canadas and Min [52] derived the impedance equation by solving the transient heat equation in the frequency domain under symmetric adiabatic conditions at both sides of the TE system. In this work, they assumed linear responses to the AC currents, i.e. no Joule heating, which is a quadratic function of current ($\sim I^2$). This assumption is valid if a relatively small current is used so that the Peltier effects at both sides are dominant over the Joule heating. In their model, they quantified the effects of contacts in the impedance spectrum, which becomes significant at high frequencies.

Figure 8.9 shows a typical impedance response in the complex coordinates from a TE element under a symmetric adiabatic condition [52]. With contributions from the contacts, the response almost becomes a semi-circle (Fig. 8.9). By measuring the low-frequency-limit resistance, $R_S + R_{AC}$ (x-intercept at $f \to 0$) and the high-frequency x-intercept, R_{AC}, as shown in the figure, and by simply taking the ratio between the two resistances similarly as in the Harman method, the figure of merit of the material can be directly obtained as in

$$ZT = \frac{R_S}{R_{AC}} \tag{8.21}$$

Physically, the difference between the low-f-limit resistance and high-f one, i.e. $R_S = V_S/I$, is the resistance accounting for the Seebeck voltage (V_S) created due to the development of temperature

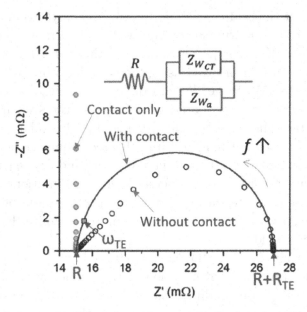

Fig. 8.9: An example of impedance response in the complex coordinate with varying AC signal frequency showing with and without contribution from contacts.

Source: Reproduced with permission from *J. Appl. Phys.*, **116**, 174510 (2014).

gradient in the system by the Peltier effect at the input current magnitude I under the adiabatic conditions, so that $R_S =^2 TL/(\kappa A)$, where T is the middle temperature, L and A are, respectively, the length and cross-sectional area of the TE element. R_{AC} is the resistance of the system purely by the ohmic response, such that $R = L/\sigma A$. The characteristic frequency ω_{TE} is additionally obtained from the impedance spectrum, which gives the thermal diffusivity of the material as

$$\omega_{TE} = \frac{4\alpha_{TE}}{L^2} \tag{8.22}$$

If the heat capacity is known, the thermal conductivity κ is found along with the thermal diffusivity extracted from ω_{TE}. Then, the Seebeck coefficient S is obtained from R_{TE}, and the electrical conductivity σ is obtained from R, completing the full characterization of all the individual properties.

More recently, Thiébaut *et al.* [53] solved the heat equations with accounting for the nonlinear effects from the Joule heating and the Peltier effects, both of which raise the second harmonic responses at 2ω. They also solved for different boundary conditions, e.g. with isothermal boundary condition at one side (heatsink) and adiabatic condition at the other side. They found that the thermal conductivity can be directly measured, without prior knowledge of the heat capacity, by measuring both the first harmonic and second harmonic responses.

8.6 Summary

In this chapter, we have discussed the several key techniques used for individual material characterization, and device characterization with capabilities to directly extract the material figure of merit. Electrical measurements for the electrical conductivity and Seebeck coefficient measurement include several four-probe methods, and non-contact remote sensing techniques based on the eddy current analysis. Thermal conductivity measurement is performed by both steady-state and transient thermal response to controlled external heat sources. Careful analysis on the parasitic heat losses is a key to the accurate characterization with these techniques. Recently, nano/micro-scale thermal characterization methods have been developed thanks to the advances in nanoscale and atomic scale probing technologies and optical thermography. Thermoreflectance imaging technique is useful in that a fast 2D mapping of surface temperature can be accurately achieved for actual devices under bias in ultra-fast temporal resolution. The Harman method and impedance spectroscopy are widely used for direct measurement of the figure of merit. Depending on the sample dimensions and surrounding conditions, an appropriate technique must be chosen for accurate device and material characterization.

References

[1] J.-H. Bahk, T. Favaloro and A. Shakouri, Thin film thermoelectric characterization techniques, *Ann. Rev. Heat Trans.*, **16**, 51–99 (2013), G. Chen (Ed.), *Annual Reviews of Heat Transfer*, New York: Begell House Inc.

[2] A. Nour Eddine *et al.*, Modeling and simulation of a thermoelectric generator using bismuth telluride for waste heat recovery in automotive diesel engines, *J. Electron. Mater.*, **48**, 2036–2045 (2019).

[3] P. M. Mayer, D. Lüerssen, R. J. Ram and J. Hudgings, Theoretical and experimental investigation of the thermal resolution and dynamic range of CCD-based thermoreflectance imaging, *J. Opt. Soc. Am. A. Opt. Image Sci. Vis.*, **24**, 1156 (2007).

[4] B. Vermeersch, J.-H. Bahk, J. Christofferson and A. Shakouri, Thermoreflectance imaging of sub 100 ns pulsed cooling in high-speed thermoelectric microcoolers, *J. Appl. Phys.*, **113**, 104502 (2013).

[5] T. C. Harman, Special techniques for measurement of thermoelectric properties, *J. Appl. Phys.*, **29**, 1373 (1958).

[6] E. Barsoukov and J. R. Macdonald, Eds., *Impedance Spectroscopy: Theory, Experiment, and Applications*, 3rd ed.: Wiley (2018).

[7] B. Yang, J. L. Liu, K. L. Wang and G. Chen, Simultaneous measurements of seebeck coefficient and thermal conductivity across superlattice, *Appl. Phys. Lett.*, **80**, 1758 (2002).

[8] M. Murakami, A. Tatami and M. Tachibana, Fabrication of high quality and large area graphite thinfilms by pyrolysis and graphitization of polyimides, *Carbon*, **145**, 23–30 (2019).

[9] A. A. Balandin, Thermal properties of graphene and nanostructured carbon materials, *Nat. Mater.*, **10**, 569–581 (2011).

[10] A. A. Ramadan, R. D. Gould and A. Ashour, On the van der pauw method of resistivity measurements, *Thin Sol. Fil.*, **239**, 272–275 (1994).

[11] L. J. Van der Pauw, A method of measuring specific resistivity and Hall effect of discs of arbitrary shape, *Philips Res. Rep.*, **13**, 1–9 (1958).

[12] Am. Soc. Test. Mat., Standard method for measuring resistivity of silicon slices with a collinear four-point probe, in *1996 Annual Book of ASTM Standards*. West Conshohocken, PA (1996).

[13] M. P. Albert and J. F. Combs, Correction factors for radial resistivity gradient evaluation of semiconductor slices, *IEEE Trans. Electron Dev.*, ED-**11**, 148 (1964).

[14] R. Rymaszewski, Relationship between the correction factor of the four-point probe value and the selection of potential and current Electrodes, *J. Sci. Instrum.*, **2**, 170 (1969).

[15] J. Krupka, Contactless methods of conductivity and sheet resistance measurement for semiconductors, conductors and superconductors, *Meas. Sci. Technol.*, **24**, 062001 (2013).

[16] R. Stibal, J. Windscheif and W. Jantz, Contactless evaluation of semi-insulating GaAs wafer resistivity using the time-dependent charge measurement, *Semicond. Sci.*, **6**, 955–1001 (1991).

[17] G. L. Miller, D. A. H. Robinson and J. D. Wiley, Contactless measurement of semiconductor conductivity by radio-frequency-free-carrier power absorption, *Rev. Sci. Instrum.*, **47**, 799–805 (1976).

[18] C. Dames, Measuring the thermal conductivity of thin films: 3 Omega and related electrothermal methods [Chapter 2], in *Annual Review of Heat Transfer*, Vol. 16, G. Chen (Ed.). Begell House (2013).

[19] D. Zhao, X. Qian, X. Gu, S. A. Jajja and R. Yang, Measurement techniques for thermal conductivity and interfacial thermal conductance of bulk and thin film materials, *J. of Elec. Packa.*, **138**, 040802 (2016).

[20] W. J. Parker, R. J. Jenkins, C. P. Butler and G. L. Abbott, Flash method of determining thermal diffusivity, heat capacity, and thermal conductivity, *J. Appl. Phys.*, **32**, 1679 (1961).

[21] P. Gill, T. T. Moghadam and B. Ranjbar, Differential scanning calorimetry techniques: Applications in biology and nanoscience, *J. Biomol. Tech.*, **21**, 167–193 (2010).

[22] D. G. Cahill, Thermal conductivity measurement from 30 to 750 K: The 3-omega method, *Rev. Sci. Instrum.*, **61**, 802–808 (1990).

[23] T. Borca-Tasciuc, A. R. Kumar and G. Chen, Data reduction in 3-omega method for thin-film thermal conductivity determination, *Rev. Sci. Instrum.*, **72**, 2139–2147 (2001).

[24] C. A. Paddock and G. A. Eesley, Transient thermoreflectance from thin metal films, *J. Appl. Phys.*, **60**, 285 (1986).

[25] D. G. Cahill, Analysis of heat flow in layered structures for time-domain thermoreflectance, *Rev. Sci. Instrum.*, **75**, 5119 (2004).

[26] A. J. Schmidt, Pump-Probe thermoreflectance, in *Annu. Rev. Heat Transfer*, G. Chen, Ed.: Begell House, ch. 6, 159–181 (2013).

[27] M. Kuball *et al.*, Measurement of temperature in active high-power AlGaN/GaN HFETs using Raman spectroscopy, *IEEE Elect. Dev. Lett.*, **23**, 7–9 (2002).

[28] S. Choi, E. R. Heller, D. Dorsey, R. Vetury and S. Graham, Thermometry of AlGaN/GaN HEMTs using multispectral Raman Features, *IEEE Transac. Elect. Dev.*, **60**, 6, 1898–1904 (2013).

[29] M. Kuball and J. W. Pomeroy, A Review of Raman thermography for electronic and opto-electronic device measurement with Sub-micron spatial and nanosecond temporal resolution, *IEEE Transac. on Dev. and Mat. Reli.*, **16**, 4, 667–684 (2016).

[30] N. Lundt *et al.*, High spatial resolution Raman thermometry analysis of TiO_2 microparticles, *Rev. Sci. Instrum.*, **84**, 104906 (2013).

[31] M. Nonnenmacher and H. K. Wickramasinghe, Scanning probe microscopy of thermal conductivity and subsurface properties, *Appl. Phys. Lett.*, **61**, 168 (1992).

[32] A. Majumdar, J. P. Carrejo and J. Lai, Thermal imaging using the atomic force microscope, *Appl. Phys. Lett.*, **62**, 2501 (1993).

[33] M. Maywald, R. J. Pylkki and L. J. Balk, Imaging of local thermal and electrical conductivity with scanning force microscopy, *Scan. Micr.*, **8**, 181–188 (1994).

[34] F. Menges, H. Riel, A. Stemmer and B. Gotsmann, Nanoscale thermometry by scanning thermal microscopy, *Rev. Sci. Instrum.*, **87**, 074902 (2016).

[35] J. Varesi and A. Majumdar, Scanning Joule expansion microscopy at nanometer scales, *Appl. Phys. Lett.*, **72**, 37 (1998).

[36] K. L. Grosse, M.-H. Bae, F. Lian, E. Pop and W. P. King, Nanoscale Joule heating, Peltier cooling and current crowding at graphene-metal contacts, *Nat. Nanote.*, **6**, 287–290 (2011).

[37] A. Lewis and K. Lieberman, Near-field optical imaging with a non-evanescently excited high-brightness light source of sub-wavelength dimensions, *Nature*, **354**, 214–216 (1991).

[38] E. Betzig, J. K. Trautman, T. D. Harris, J. S. Weiner and R. L. Kostelak, Breaking the diffraction barrier: optical microscopy on a nanometric scale, *Science*, **251**, 1468 (1991).

[39] A. Majumdar, Scanning thermal microscopy, *Ann. Rev. of Mate. Sci.*, **29**, 505–585 (1999).

[40] M. Vollmer and K.-P. Möllmann, *Infrared Thermal Imaging: Fundamentals, Rese. and Appl.*, 2nd ed.: Wiley (2017).

[41] S. Dhokkar, B. Serio, P. Lagonotte and P. Meyrueis, Power transistor near-infrared microthermography using an intensified CCD camera and frame integration, *Meas. Sci. and Tech.*, **18**, 2696 (2007).

[42] O. Breitenstein, F. Altmann, T. Riediger, D. Karg and V. Gottschalk, Lock-in thermal IR imaging using a solid immersion lens, *Microel. Reli.*, **46**, 1508–1513 (2006).

[43] O. Breitenstein *et al.*, Microscopic lock-in thermography investigation of leakage sites in integrated circuits, *Rev. Sci. Instrum.*, **71**, 4155(2000).

[44] H. Köck, V. Kosel, C. Djelassi, M. Glavanovics and D. Pogany, IR thermography and FEM simulation analysis of on-chip temperature during thermal-cycling power-metal reliability testing using in situ heated structures, *Microel. Reli.*, **49**, 1132–1136 (2009).

[45] T. Favaloro, J.-H. Bahk and A. Shakouri, Characterization of the temperature dependence of the thermoreflectance coefficient for conductive thin films, *Rev. Sci. Instrum.*, **86**, 024903 (2015).

[46] K. Maize, J. Christofferson and A. Shakouri, Transient thermal imaging using thermoreflectance, in *The 24th Annual IEEE Semiconductor Thermal Measurement and Management Symposium*, San Jose, CA, USA, 55–58 (2008).

[47] K. Maize *et al.*, High resolution thermal characterization and simulation of power AlGaN/GaN HEMTs using Micro-Raman thermography and 800 picosecond transient thermoreflectance imaging, in *IEEE Compound Semiconductor Integrated Circuit Symposium (CSICs)*, 1–8 (2014).

[48] D. Kendig, K. Yazawa and A. Shakouri, Transient thermal measurement and behavior of integrated circuits, in *29th IEEE Semiconductor Thermal Measurement and Management Symposium*, 206–210 (2013).

[49] K. Yazawa, D. Kendig and A. Shakouri, Sub-Micron thermal imaging characterization of GaN HEMT and MMIC devices, in *2015 IEEE Compound Semiconductor Integrated Circuit Symposium (CSICS)*, 1–4 (2015).

[50] K. Maize *et al.*, Super-Joule heating in graphene and silver nanowire network, *Appl. Phys. Lett.*, **106**, 143104 (2015).

[51] M. A. Wahab, S. H. Shin and M. A. Alam, 3D Modeling of spatio-temporal Heat-transport in III-V Gate-all-around transistors allows accurate estimation and optimization of nanowire temperature, *IEEE Trans. Elect. Dev.*, **62**, 3595 (2015).

[52] B. Beltrán-Pitarch, J. Prado-Gonjal, A. V. Powell and J. García-Cañadas, Experimental conditions required for accurate measurements of electrical resistivity, thermal conductivity, and dimensionless figure of merit (ZT) using Harman and impedance spectroscopy methods, *J. Appl. Phys.*, **125**, 025111 (2019).

[53] W. Kobayashi, W. Tamura and I. Terasaki, Thermal conductivity and dimensionless figure of merit of thermoelectric rhodium oxides measured by a modified Harman method, *J. Electron. Mater.*, **38**, 964–967 (2009).

[54] A. D. Downey, T. P. Hogan and B. Cook, Characterization of thermoelectric elements and devices by impedance spectroscopy, *Rev. Sci. Instrum.*, **78**, 093904 (2007).

[55] J. García-Cañadas and G. Min, Impedance spectroscopy models for the complete characterization of thermoelectric, *J. Appl. Phys.*, **116**, 174510 (2014).

[56] E. Thiébaut, F. Pesty, C. Goupil, G. Guegan and P. Lecoeur, Non-linear impedance spectroscopy applied to thermoelectric measurements: Beyond the ZT estimation, *arXiv:1807.01387v2* (2018).

[57] S. Dilhaire *et al.*, Determination of zT of p-n thermoelectric couples by AC electrical measurement, in *Proceedings 21st Inter. Conf. Thermoelectrics*, 321–324 (2002).

[58] K. E. Hnida *et al.*, Tuning of the seebeck coefficient and the electrical and thermal conductivity of hybrid materials based on polypyrrole and bismuth nanowires, *Chem. Phys. Chem.*, **19**, 1617–1626 (2018).

Chapter 9

Simulation Tools

In this chapter, several online simulation tools for thermoelectric devices and materials are introduced. From thin-film devices to multi-element modules and systems with cost-performance trade-off analysis, this simulation package covers a variety of thermoelectric systems for both power generation and cooling applications. Thermoelectric performances can be calculated as a function of various design parameters in these simulation tools to find optimal design conditions and learn the basic principles. For material simulation, a Boltzmann transport simulator has been developed, which offers prediction of thermoelectric material properties based on the band structure and scattering parameters of the material. Finally, we introduce an online course developed for education on topics in thermoelectrics from the principles and atomic-scale devices to large-scale systems and applications. In addition to the wide use of the simulation tools expected for individual research of engineers and scientists, the online course is also designed to utilize the tools for students learning with associated exercise problems.

9.1 Introduction

Thermoelectric principles are based on the fact that the electrical conduction and the heat conduction in a material are mutually coupled. The Seebeck coefficient, also called as the thermopower, is the fundamental property that connects these two transport phenomena being defined as the ratio between the correlated electric potential (ΔV) and thermal potential (ΔT) such that $S = -\Delta V/\Delta T$.

The electrical conduction is governed by the Ohmic law, and the heat conduction is governed by the Fourier law. By including the Seebeck effect that links the two physical laws, the coupled charge and heat equations are found to be [1]

$$\vec{J} = \sigma\vec{E} - \sigma S\nabla T \qquad (9.1)$$

$$\vec{q} = ST\vec{J} - \kappa\nabla T \qquad (9.2)$$

where $\vec{J}$ and $\vec{q}$ are, respectively, the electric and heat current densities, σ is the electrical conductivity, κ is the thermal conductivity, and ∇T is the temperature gradient. With a non-zero Seebeck coefficient ($S \neq 0$), the second term on the right side of Eq. (9.1) and the first term of Eq. (9.2) link the two equations. With a zero Seebeck coefficient ($S = 0$), Eq. (9.1) becomes the Ohmic equation, and Eq. (9.2) becomes the Fourier equation. The first term of the right side of Eq. (9.2) is the Peltier effect term that depicts the amount of heat flux carried by the electric flux, and ST in the term is also known as the Peltier coefficient, i.e. $\Pi = ST$.

Due to this coupling between the electrical and heat conductions, one must solve the two coupled equations together to quantify the net thermoelectric effects, which is often a difficult task. Also, a thermoelectric device is in a complex form involving multiple elements of p-type and n-type semiconductors and temperature-dependent and/or location-dependent material properties of them. Hence, a numerical simulation is beneficial for the fast and accurate prediction and optimization of thermoelectric performance.

In this chapter, we introduce four simulation tools that are available online at nanoHUB.org for public use. The first tool is called 'Thin-film and Multi-element Thermoelectric Devices Simulator' or THERMO tool [2], which can simulate both thin-film devices and multi-element modules for both power generation and cooling applications with constant material properties. The second tool is named 'Advanced Thermoelectric Power Generation Simulator for Waste Heat Recovery and Energy Harvesting' or ADVTE tool [3], which is useful to simulate multi-element modules with temperature-dependent material properties. The third simulation tool is 'Thermoelectric System Performance Optimization and Cost Analysis Simulator' or TEDEV tool [4], which can perform co-optimization of module performance and cost for thermoelectric systems. The last tool is called 'Linearized Boltzmann Transport calculator for Thermoelectric Materials' or BTE solver tool [5], which is used for material simulations based on the linearized Boltzmann transport equations (BTE) and non-parabolic band models. Users need to sign in on nanoHUB.org to run the tools on their web browsers.

An online course [6] has been developed on the nanoHUB-U platform for open education and research on thermoelectrics (Fig. 9.1). Through this course, students and learners can watch lecture videos and solve exercise problems online using the simulation tools to learn a broad range of topics in thermoelectrics in their own learning paces. In the following sections, we will discuss each of the simulation tools and the online course in more details.

9.2 Thin-film and Multi-element TE Device Simulator (THERMO Tool)

The thin-film and multi-element device simulator or the THERMO tool [2] is capable of simulating thermoelectric devices made of thin-films or multiple thermoelectric elements for both power generation and cooling applications in steady state under various boundary conditions. The schematic of thermoelectric devices that can be simulated by THERMO tool is shown in Fig. 9.2.

9.2.1 *Simulation of thin-film TE devices*

A typical thin-film device simulated in THERMO tool is schematically shown in Fig. 9.2(a). The thermoelectric film is patterned within a small cross-section for current and heat confinement. Since it is too thin to stand alone or be carried alone, the film is deposited on a substrate of a much larger size. A conductive substrate is useful to conveniently make a ground contact at the bottom, so that both the heat and electrical current flow through both the film and the substrate. THERMO tool is able to account for the substrate effects, including

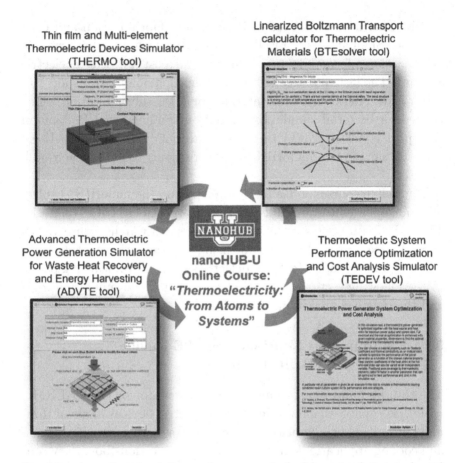

Fig. 9.1:   An online simulation package comprised of four simulation tools and an online course for research and education on thermoelectricity from atoms to systems.

the current spreading in the substrate and the Peltier effects at the interface. A top contact is made on top of the film, which is connected to a side electrode for easy probing. In order to prevent the side-wall current leakage and ensure a vertical current flow, an insulation layer must be deposited on the sidewall of the film, after which the top contact and side electrode are deposited and patterned by lithography.

There are total four modes a user can choose from for thin-film device simulation, which are the four combinations of two boundary

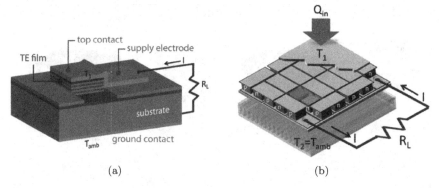

Fig. 9.2: Schematics of thermoelectric devices that can be simulated in THERMO tool. (a) Thin-film device sitting on a substrate and (b) multi-element module with ceramic plates on both sides.

conditions (constant heat flux or constant temperature at the top surface) and two applications (power generation or cooling). One can choose one of the options from the pull-down menu in the second page of the tool as shown in the screenshots in Figs. 9.3(a) and (b), which, respectively, show the cooling mode with constant cooling power boundary condition, and the power generation mode with constant temperature boundary condition. Once a mode is selected, the user can then select an independent variable associated with the mode, against which all the output values are plotted once the simulation is finished. Typical design parameters used as an independent variable are the thickness and cross-sectional area of the TE film. One can also choose one of the material properties as an independent variable.

In the next page in the tool, the user can enter the material properties and other parameters that affect performance such as the contact resistance and the substrate properties as shown in Fig. 9.3(c). By clicking the selective button adjacent to each property label, the user can manually enter the values. Note that the material properties are assumed to be constant in THERMO tool, e.g. using the average values of the properties over the temperature range applied.

We model the thin-film device as a two-layer TE element with one-dimensional heat and current flow in the vertical direction with

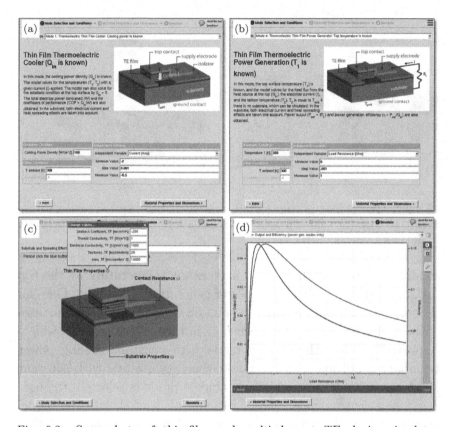

Fig. 9.3: Screenshots of thin-film and multi-element TE device simulator (THERMO tool) showing (a) selection of Mode 1 for thin-film cooling with heat input boundary condition, (b) selection of Mode 4 for thin-film power generation with temperature boundary condition, (c) inputs of material properties, and (d) a typical simulation result plot (power output and efficiency) as a function of an independent variable.

respect to the orientation shown in Fig. 9.3(a) in THERMO tool. Figure 9.4 shows the thermal and electrical circuit models used for the simulation. The two layers, thin film and substrate, are connected in series in both circuits. In the thermal circuit, Node 1 represents the top surface of the thin film, and Node 2 represents the interface between the thin film and the substrate. The bottom side of the substrate is assumed to be at a fixed temperature, the ambient temperature, T_{amb}. There are three kinds of heat flowing in or out of each

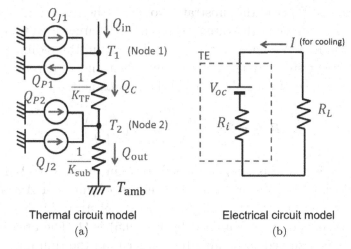

Thermal circuit model
(a)

Electrical circuit model
(b)

Fig. 9.4: (a) Thermal circuit model and (b) electrical circuit model for the thin-film devices in THERMO tool.

node: the conduction heat, Q_C, the Peltier heat, Q_P, and the Joule heat, Q_J. Another conduction heat, Q_{in}, flows into Node 1 from the top surface, which can be a heat input from the heat source in the case of power generation mode, or the cooling power absorbed from the top side in the case of cooling mode. Also, there is a conduction heat, Q_{out}, going out of Node 2 towards the substrate.

Joule heating occurs uniformly inside each layer and flows out of the layer at the two sides of the layer equally. At Node 1, assuming that the electrode has negligibly small resistance and there is no additional layer on top of the film, the Joule heat is the sum of a half of the Joule heat generated from the thin film and the Joule heat by the contact resistance. At Node 2, the Joule heat is the sum of a half of the Joule heat generated from the film and a half of the Joule heat generated from the substrate, such that

$$Q_{J1} = \frac{1}{2}I^2 R_{\text{TF}} + I^2 R_c \tag{9.3}$$

$$Q_{J2} = \frac{1}{2}I^2 R_{\text{TF}} + \frac{1}{2}I^2 R_{\text{sub}} \tag{9.4}$$

where I is the electric current, R_{TF} is the electrical resistance of the thin film, R_c is the contact resistance at Node 1, and R_{sub} is the electrical resistance of the substrate. Note that there is a large spreading of current in the substrate as it has a much larger area than that of the thin film. Vashaee *et al.* [7] found from full three-dimensional finite element simulations that the equivalent one-dimensional electrical resistance of the substrate can be approximated as

$$R_{\text{sub}} = 3570 \frac{\rho_{\text{sub}}}{\sqrt{A_{\text{TF}}}} \tanh\left(\frac{t_{\text{sub}}}{71.9}\right) \tag{9.5}$$

where ρ_{sub} is the resistivity of the substrate in the unit of Ω cm, t_{sub} is the thickness of the substrate in the unit of μm, and A_{TF} is the cross-sectional area of the thin film on the substrate in the unit of μm^2. The unit of R_{sub} obtained by Eq. (9.5) is Ω. The area of the substrate is assumed to be infinitely larger than the thin film.

Peltier effects occur at interface between two layers due to the difference in the Peltier coefficients of the two layers. In the thin-film device, Node 1 is the interface between the thin film and the top electrode. Assuming that the Peltier coefficient ($\Pi = ST$) of the top electrode is much smaller than that of the thin film, the Peltier heat at Node 1 simply becomes

$$Q_{P1} = S_{\text{TF}} T_1 I \tag{9.6}$$

At Node 2, the thin film and the substrate form the interface, and the Peltier heat is given by

$$Q_{P2} = (S_{\text{TF}} - S_{\text{sub}}) T_2 I \tag{9.7}$$

The conduction heats flowing through the thin film and the substrate are given, respectively, by

$$Q_C = K_{\text{TF}}(T_1 - T_2) \tag{9.8}$$

$$Q_{\text{out}} = K_{\text{sub}}(T_2 - T_{\text{amb}}) \tag{9.9}$$

Here, the thermal conductance of the thin film is obtained assuming a 1D heat flow through the thin film as $K_{\text{TF}} = \kappa_{\text{TF}} \frac{A_{\text{TF}}}{t_{\text{TF}}}$, and the thermal conductance of the substrate includes the 3D heat spreading, given by [8]

$$K_{\text{sub}} = \kappa_{\text{sub}} \frac{\sqrt{\pi A_{\text{TF}}}}{\tanh\left(\frac{t_{\text{sub}}}{\sqrt{A_{\text{TF}}}}\right)} \tag{9.10}$$

Finally, a heat balance equation can be formulated using these components of heat at each node, depicting that total heat input is equal to total heat output at each node in steady state. According to the thermal circuit model, the heat balance equations at Nodes 1 and 2 are found, respectively, to be

$$Q_{\text{in}} + Q_{J1} - Q_{P1} = Q_C \tag{9.11}$$

$$Q_{J2} - Q_{P2} + Q_C = Q_{\text{out}} \tag{9.12}$$

Another equation is obtained from the electrical circuit model, which is shown in Fig. 9.4(b). In this electrical circuit, the open circuit voltage, V_{oc}, is induced by the Seebeck effect in both the thin film and the substrate. The internal resistance, R_i, is also the sum of those from both layers and from the top contact, such that

$$V_{\text{oc}} = S_{\text{TF}}(T_1 - T_2) + S_{\text{sub}}(T_2 - T_{\text{amb}}) \tag{9.13}$$

$$R_i = R_{\text{TF}} + R_{\text{sub}} + R_c \tag{9.14}$$

The TE device is connected to a load resistance, R_L, in the case of power generation mode, to make a closed circuit and drive a current to the load. By the Ohmic law,

$$V_{OC} = I(R_i + R_L) \tag{9.15}$$

So far, we have formulated totally three equations, Eqs. (9.11), (9.12), and (9.15), for three unknown variables. The variables are Q_{in}, T_2, and I if T_1 is known as the constant temperature boundary condition (Mode 4 in THERMO tool). Or they are T_1, T_2, and I if Q_{in} is known as the constant heat flux boundary condition (Mode 3 in THERMO tool). Therefore, the unique solution of the three variables can be found by solving the three equations together.

The primary performance measures of a TE power generator are the power output and the efficiency. Finally, they are obtained by

$$P_{\text{out}} = IR_L \tag{9.16}$$

$$\eta = \frac{P_{\text{out}}}{Q_{\text{in}}} \tag{9.17}$$

In the cooling mode, there is no load resistance, and the current is a user input. Therefore, only the two thermal Eqs. (9.11), and (9.12), are solved together to obtain the two unknown variables. The variables are Q_{in} and T_2 if T_1 is known as the constant temperature boundary condition (Mode 2 in THERMO tool). Or they are T_1 and T_2 if Q_{in} is known as the constant heat flux boundary condition (Mode 1 in THERMO tool).

Performance measures of a TE cooler are the cooling power, the temperature at the cooled side, and the coefficient of performance (COP). In the case of constant temperature boundary condition, the cooling power, Q_{in}, is obtained by solving the heat balance equations as discussed above. In the case of constant heat flux boundary condition, of which one example is the adiabatic condition at the cooled side, i.e. $Q_{\text{in}} = 0$, the resulting temperature at the cooled side, T_1, is obtained instead. Finally, the COP is obtained by

$$\text{COP} = \frac{Q_{\text{in}}}{W} \tag{9.18}$$

where W is the electric power consumed for the cooling operation, given by

$$W = IV_{\text{oc}} + I^2 R_i \tag{9.19}$$

Note that this power consumption includes the power consumed to overcome the internal open-circuit voltage in order to drive the current I, i.e. IV_{oc}, as well as the ohmic power loss in the TE device, i.e. $I^2 R_i$. At the end of the simulation, these performance measures are plotted as a function of the independent variable that the user selected.

9.2.2 *Simulation of multi-element TE modules*

Simulation of a multi-element module is similar to the simulation in the thin-film device, except for the involvement of two types of TE elements, n-type and p-type, as shown in Fig. 9.2(b). Thus, two sets of design parameters for the two types of elements are needed and entered as user inputs. Constant material properties are assumed. In THERMO tool, the external thermal resistances by the ceramic plates and heat sinks are neglected and the boundary temperatures of the two types of elements are assumed to be the same. The bottom-side temperature of the elements is assumed to be the same as the ambient temperature. Also, there is no substrate unlike the thin film case. The ADVTE tool can account for the external thermal resistances, and many other factors that are not taken into account in THERMO tool. We will discuss the ADVTE tool later in the following section.

Several design parameters can be selected as independent variables for multi-element module simulation in THERMO tool. The cross-sectional areas of the n-type and p-type elements, and their thickness are such design parameters. The top-side temperature, load resistance (in power generation mode), and the electric current (in cooling mode) can also be selected as independent variables. THERMO tool offers only a constant temperature boundary condition for module simulation, which is that the top-side temperature, T_1, is assumed to be constant and given. Number of element pairs is a user input.

As similarly discussed in the thin film simulation, heat balance equations are formulated for steady-state analysis at the interfaces and solved to determine the operating temperature and electrical/heat currents based on the thermal and electrical circuit models. Total heat input is the sum of the heat inputs in all the elements. Total open-circuit voltage and total electrical resistance are also obtained by adding the voltages and resistances of the individual elements because the elements are connected in series electrically:

$$Q_{\text{in}} = N_{\text{pair}}(Q_{\text{in},n} + Q_{\text{in},p}) \tag{9.20}$$

$$V_{oc} = N_{pair}(V_{oc,n} + V_{oc,p}) \qquad (9.21)$$

$$R_i = N_{pair}(R_{i,n} + R_{i,p}) \qquad (9.22)$$

where N_{pair} is the number of n-type and p-type element pairs, and subscripts n and p indicate the quantity in each element of n-type and p-type, respectively. Then, the performance measures such as power output, efficiency, cooling power and COP can be obtained using the same equations used for the thin-film device simulation, Eqs. (9.16)–(9.19).

9.2.3   *Simulation of double-segmented elements*

In addition to the thin film and module simulations, THERMO tool is also capable of simulating segmented elements made of two segments of equal cross-sectional area. One can use the thin-film layer as the first segment and the substrate layer as the second. In order to keep the cross-sectional areas of the two segments the same, the spreading effect in the substrate must be turned off. The user can choose the option 'Substrate with No Spreading Effect', in the pull-down menu in the third page of THERMO tool to remove the spreading effect as shown in Fig. 9.5. Then use can input the material properties and dimensions (thickness) for the two segments.

Bian and Shakouri [9] pointed out that a double-segmented element can perform a better cooling than a single-segmented element can do at their maximum cooling conditions, if the interface between the two segments is carefully designed in such a way the Peltier cooling heat at the interface completely cancels the total Joule heating coming into the interface. In other words, Eq. (9.23) needs to be satisfied at the interface between the two segments:

$$Q_{P2} = Q_{J1} + Q_{J2} \qquad (9.23)$$

where $Q_{P2} = (S_2 - S_1)T_2 I$ is the Peltier cooling heat at the interface, Q_{J1} is the half of the Joule heating from the first segment that comes out to the interface, and Q_{J2} is the half of the Joule heating from the second segment that comes out to that interface. The right-hand side of Eq. (9.23) is the total Joule heating coming into the interface. Note that for cooling at the top surface, $(S_2 - S_1)$ must be positive,

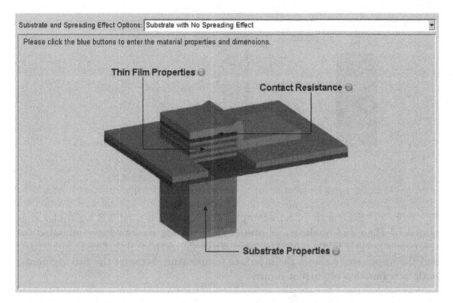

Fig. 9.5: Simulation of a double-segmented TE leg in THERMO tool with the option of 'Substrate with No Spreading Effect'.

i.e. $S_2 > S_1$ for both p-type segments $(S_1, S_2 > 0)$ or $|S_1| > |S_2|$ for both n-type segments $(S_1, S_2 < 0)$.

One example cooler design was suggested by Bian and Shakouri [9], in which the material for the second segment possesses a twice large Seebeck coefficient that that of the first material, while the power factors and the electrical resistances of the two segments are kept the same, i.e. $S_2 = 2S_1$, $\sigma_2 = \sigma_1/4$, and $R_2 = R_1$. To meet the last condition for resistance, the thickness of the second segment is chosen to be one quarter of the thickness of the first segment to compensate the four times smaller electrical conductivity, i.e. $t_2 = t_1/4$. This cooler design example is schematically shown in Fig. 9.6(a).

THERMO tool can simulate this example cooler design made of two segments with the 'no spreading effect' option as discussed above. Figure 9.6(b) displays the simulation result from THERMO tool, showing the cooling performance or the top surface temperature of the double-segmented cooler as a function of current input. As shown in the figure, the double segment cooler can achieve a large cooling

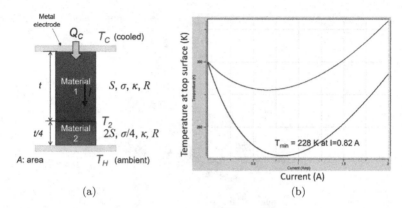

Current (A)

(a)   (b)

Fig. 9.6: (a) Schematic of an example of double-segmented TE leg cooler suggested by Bian and Shakouri [9] and (b) the cooling performance simulated for this cooler by THERMO tool. The lower curve shows the top surface temperature, and the upper curve shows the interface temperature between the two segments, both as a function of current input.

to T_C = 228 K or ΔT = 72 K with ambient temperature 300 K at the top surface at I = 0.82 A, which is about 40% enhancement compared to the maximum cooling ΔT = 52 K that can be achieved by a single segment of the same first material.

9.3   Advanced Thermoelectric Power Generation Simulator (ADVTE tool)

The Advanced Thermoelectric Power Generation Simulator, also known as ADVTE tool [3], is an upgraded version of THERMO tool for TE module simulation. ADVTE tool is capable of simulating more realistic thermoelectric power generation performance of multi-element TE modules particularly for waste heat recovery application, where a large temperature difference is typically applied across the module. THERMO tool only accepts constant material properties for simulation, so that it cannot accommodate the temperature-dependency of material properties, which is important for realistic TE simulations for those applications. Furthermore, finite external heat transfer coefficients at both sides and thermal conductivity of gap filler material can be taken into account in ADVTE tool, which was not possible in THERMO tool.

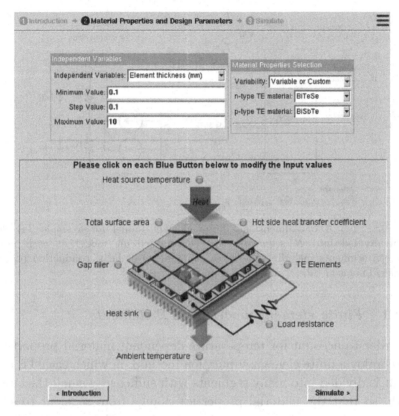

Fig. 9.7: Screenshot of ADVTE tool that shows input boxes, menus, and buttons for user input and option choice.

Figure 9.7 shows a screen shot of the second page of ADVTE tool, where the user can input the design parameters and material properties and select an independent variable for which the performance will be simulated. Here, the user can select the options for material properties, which can be either constant, or variable/custom as a function of temperature. In the variable/custom option, the user can select the materials from the built-in database, which include Bi_2Te_3-based alloys and SiGe for both n-type and p-type, and $Mg_2(SnSi)$ for n-type, and PbTe alloys for p-type elements. Or, the user can upload their own material properties saved in the format of the template file and then run the simulation with their own materials.

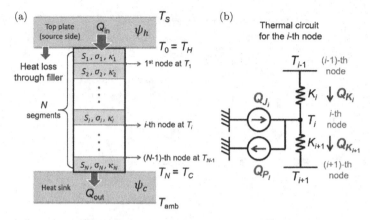

Fig. 9.8: (a) A one-dimensional finite element model for a single TE element divided into N segments. (b) Thermal circuit model at each node between segments including the three components of heat: conduction (Q_{Ki}), Peltier/Thomson (Q_{Pi}), and Joule term (Q_{Ji}).

9.3.1 *Finite element model for TE elements*

In order to account for temperature-dependent material properties, we employ a finite element simulation method, in which each TE element is divided into many segments with sufficiently small thickness each, so that the material properties change from segment to segment. In each segment, we assume constant material properties. This is a valid assumption for infinitesimally thin layers across in which the temperature difference applied is small enough to ignore. The number of segments depends on how large the total temperature difference is applied across the module, but we typically use more than 100 segments for each element. Figure 9.8 displays the finite element model for a TE element with N segments, and the thermal circuit model at each node between adjacent segments, e.g. i-th node.

For a multi-element module, we use two major design parameters as independent variables, the fill factor F and the thickness of TE element t_{TE}. First, the fill factor is defined as the fractional area coverage of the TE elements for the module area given by

$$F = \frac{N_{\text{pair}}(A_n + A_p)}{A_{\text{total}}} \tag{9.24}$$

where N_{pair} is the number of n-type and p-type element pairs, and A_n and A_p are, respectively, the cross-sectional area of n-type and p-type elements. The second parameter is the thickness of TE elements t_{TE}. We use the same thickness for both n-type and p-type elements. By adjusting the two parameters, L_{TE} and F, independently, both the total thermal resistance and electrical resistance can be tuned independently and broadly to find out the optimal design for maximum performance.

In our finite element model, each segment has constant material properties, S_i, σ_i, and κ_i, assigned, as shown in Fig. 9.8(a). These material properties are evaluated at the average temperature across the segment, i.e. $(T_{i-1}+T_i)/2$. Since the temperature in each segment is not known *a priori*, we make a guess for the initial temperature profile, and then assign the material properties in each segment corresponding to the temperature at each position. By solving all the heat balance equations formulated at every node simultaneously, one can obtain an updated temperature profile across the TE element. Then a new set of material properties is allotted at each node based on the updated temperatures for the next iteration. This process is repeated until the temperature profile converges within an acceptable tolerance. A good initial guess for the temperature profile can be a linear temperature distribution inside TE elements, assuming that the conduction term is dominant over the Joule and Peltier heats.

At one point of iteration, the Joule heat, the Peltier heat at the i-th node ($0 \leq i \leq n$), and the conduction heat across the i-th segment are, respectively, given by

$$Q_{J_i} = \frac{1}{2}I^2 R_i + \frac{1}{2}I^2 R_{i+1} \tag{9.25}$$

$$Q_{P_i} = (S_{i+1} - S_i)T_i I \tag{9.26}$$

$$Q_{K_i} = (T_{i-1} - T_i)K_i \tag{9.27}$$

where I is the electric current, and $R_i = L_{TE}/(\sigma_i A_{TE}n)$, and $K_i = (\kappa_i A_{TE}n)/L_{TE}$, are, respectively, the electrical resistance and thermal conductance of the i-th segment. A_{TE} is the cross-sectional area of the individual TE elements, i.e. $A_{TE} = A_n$ for n-type elements, and $A_{TE} = A_p$ for p-type elements. The corresponding thermal resistance network model is shown in Fig. 9.8(b) for the i-th node. The heat balance equation at the i-th node is formulated as

$$Q_{J_i} - Q_{P_i} + Q_{K_i} - Q_{K_{i+1}} = 0 \tag{9.28}$$

where the index i varies from 1 to $(n-1)$ creating a total $(n-1)$ equations to solve. It is noted that the Peltier term between segments is non-zero if the Seebeck coefficient varies with temperature. This is also called the *Thomson effect* [10].

At the 0-th node or at the hot side of TE element, the heat balance equation has the same form as Eq. (9.28), but with $i = 0$. At this node, the Joule and Peltier terms have only one-side component because there is no Joule or Peltier effects from the top hot plate. Thus,

$$Q_{J_0} = \frac{1}{2}I^2 R_1 \tag{9.29}$$

$$Q_{P_0} = S_1 T_0 I \tag{9.30}$$

where $T_0 = T_H$. The outgoing conduction term Q_{K_1} has the same form as in Eq. (9.27) but with $i = 1$, and the incoming conduction term Q_{K_0} from the top plate at the 0-th node is determined by the heat transfer coefficient of the top plate h_h and also the heat loss through the gap filler must be taken into account, such that, respectively,

$$Q_{K_1} = (T_0 - T_1)K_1 \tag{9.31}$$

$$Q_{K_0} = \frac{1}{F}A_{TE}h_h(T_{\text{top}} - T_0)$$

$$- \frac{(1-F)}{F}\frac{A_{TE}}{L_{TE}}k_{\text{filler}}(T_0 - T_n) \tag{9.32}$$

where k_{filler} is the thermal conductivity of the gap filler in the TE module. In Eq. (9.32), the first term on the right-hand side of the equation describes the heat input to the TE element from the heat source. The second term is the heat lost through the filler by conduction, which is subtracted from the first term to get the net heat input into the TE element. The total heat input per element is simply the first term on the right side of Eq. (9.32), hence

$$Q_{\text{in}} = \frac{1}{F} A_{TE} h_h (T_S - T_H) \tag{9.33}$$

Note that the heat input to a n-type element can be different from that to a p-type element, i.e. $Q_{\text{in},n} \neq Q_{\text{in},p}$, as their hot side temperatures can be different.

Similarly, the heat balance equation at the n-th node, or at the cold side of individual TE elements can be formulated using Eq. (9.28). The Joule and Peltier terms here have only one-side component from the n-th segment of the TE element because there is no Joule or Peltier effects from the bottom cold plate. Thus,

$$Q_{J_n} = \frac{1}{2} I^2 R_n \tag{9.34}$$

$$Q_{P_n} = S_n T_n I \tag{9.35}$$

where $T_n = T_C$. The incoming conduction term to the n-th node Q_{K_n} has the same form as Eq. (9.28) but with $i = n$, and the outgoing conduction term $Q_{K_{N+1}}$ from the n-th node towards the cold plate is determined by the heat transfer coefficient of the cold plate h_C, such that, respectively,

$$Q_{K_n} = (T_{n-1} - T_n) K_n \tag{9.36}$$

$$Q_{K_{n+1}} = A_{TE} h_C (T_n - T_{\text{bot}}) \tag{9.37}$$

where T_{bot} is assumed to be equal to the ambient temperature, i.e. $T_{\text{bot}} = T_{\text{amb}}$.

Thus far, we have constructed the $(n+1)$ number of heat balance equations given by Eq. (9.28) at the $(n+1)$ number of nodes with the index i varying from 0 to n for a TE element. Since there are the same number of unknown temperatures, i.e. $T_i (i = 0 - n)$, a unique solution for the temperature profile can be found that satisfies all the equations for a given electric current I.

9.3.2 *Evaluation of TE performance*

The electric current I flowing through the module is determined from the analysis of the electrical circuit model with the load resistance R_L. Since all the elements, electrodes, and the load resistance are connected electrically in series, adding all the resistances from them along with the contact resistances results in the total internal electrical resistance in the module given by

$$R_{\text{int}} = \sum \left(\sum_i R_i + 2R_c + 2R_{\text{el}} \right) \qquad (9.38)$$

where $\sum_i R_i$ is the internal resistance of each TE element obtained by adding the resistances from all the segments of the element. R_c is the contact resistance per one side of TE element, and R_{el} is the electrode resistance per one side of TE element. The Σ indicates the summation over all the TE elements. In the module, the open circuit voltage is created by the Seebeck effects inside the TE elements. In each segment of a TE element, the Seebeck voltage is obtained as

$$V_{\text{OC}_i} = S_i(T_i - T_{i-1}) \qquad (9.39)$$

Then, the total open circuit voltage is obtained by adding all the Seebeck voltages over all the segments of all the TE elements, thus by two-level summation

$$V_{\text{OC}} = \sum \sum_i V_{\text{OC},i} \qquad (9.40)$$

Finally, the electric current I is obtained by the Ohm's law for the electric circuit model as

$$I = \frac{V_{\text{OC}}}{R_{\text{int}} + R_L} \qquad (9.41)$$

As discussed earlier, the found temperature values are re-entered to update the material properties in each segment, and this process is iterated until the temperature profile converges.

Finally, the module performances are calculated from the converged temperature profile and the electric current. The voltage output, power output, total heat input to TE, the TE efficiency are obtained, respectively, by

$$V_{\text{out}} = IR_L \tag{9.42}$$

$$P_{\text{out}} = IV_{\text{out}} \tag{9.43}$$

$$Q_{\text{in,TE}} = N_{\text{pair}}(Q_{\text{in},n} + Q_{\text{in},p}) \tag{9.44}$$

$$\eta = \frac{P_{\text{out}}}{Q_{\text{in,TE}}} \tag{9.45}$$

9.4 System Performance Optimization and Cost Analysis Simulator (TEDEV Tool)

The Thermoelectric Power Generator System Optimization and Cost Analysis, also known as TEDEV tool [4], is a tool for optimizing the design of thermoelectric power generator in conjunction with thermal performance of heat sinks. In particular, for waste heat recovery, system design for maximum power output and cost effectiveness are taken into account in this tool. Full electrical and thermal co-optimization is performed by the tool with given material properties and dimensions by finding the optimal thickness of thermoelectric elements. One can choose a material property such as Seebeck coefficient and thermal conductivity as an independent variable to optimize the performance of the power generator as a function of the chosen material property. Heat transfer coefficients of the heat sinks at the hot and cold sides can also vary as independent variables. Fractional area coverage by thermoelectric elements (fill factor) is another parameter that can be optimized for the best performance and cost in this simulation tool. A set of parameters is given as default values in TEDEV tool for an example case of thermoelectric topping co-generation steam turbine system.

Figure 9.9 shows a screenshot of the second page of TEDEV tool, where the user can input the material properties and heat sink performance of both hot and cold side. As the user goes forward to the third page, he/she can specify the detailed design specification of the module and material market prices. The cost of heat sinks are given with default values, but the user can change them to reasonable values for the application of interest.

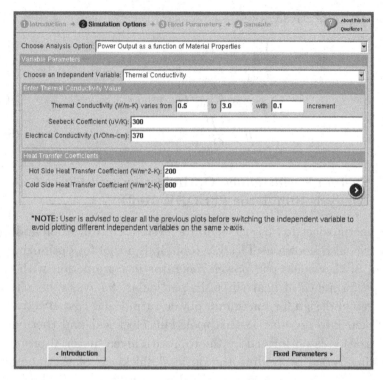

Fig. 9.9:   Screenshot of TEDEV tool that shows input boxes, menus, and buttons for user input and option choice.

9.4.1   *System performance*

Using the TEDEV tool, the system performance is expected to be optimized for maximum power output. At the same time, cost per power output can be minimized. TEDEV tool solves the hot side and cold side temperatures in accordance with the given parameters. Then, the electrical power output is obtained at the electrically optimum load (impedance matching with the factor of $\sqrt{1 + Z\bar{T}}$). See Section 5.2 for the modeling detail for the design optimization. This results in some change in the temperatures at both sides of the TE module from the initial values. The equations remain recursive due to the electro-thermal coupling until reaching the steady-state equilibrium. At the equilibrium within the allowable residue of iteration, the temperature conditions of the TE legs are determined and the power output follows. In this code, one of the thermoelectric material properties can take a range for variational study. Therefore, the user can analyze the impact of material properties for future material development or estimate what-if performance and cost with one unknown value, e.g. thermal conductivity. By connecting the Boltzmann transport equation solver module as explained in the following section, material researcher can predict the impact of a parameter to determine the band structure, e.g. effective mass, all the way to the end user benefit in cost for power.

9.4.2 Cost analysis

The generator cost consists of two major factors: material cost and assembly process cost. Material mass-based cost is known as the minimum cost limit of manufacturing as the extreme end of learning curve, where production works are infinitely effective. In reality, the process cost matters, but can be small as manufacturing technology becomes mature. Here, the code suggests the cost limit based on the material usage, but the user can still include the process cost, e.g. soldering and assembling with heat sinks. The key cost driver for thermoelectric is the dimension of TE elements (legs). When the thickness d is optimized for maximum power output, it is determined by external thermal resistances of heat sinks as well as fractional area coverage (fill factor) F. Impact of the fill factor is large since the volume of the leg is also proportional to cross sectional area A, hence *mass* $\propto F^2$ for the legs. Reminding the maximum power design, the optimum thickness is described as follows: (see Section 5.2 for details.)

$$d_{opt} = \frac{4}{\alpha^2} m\beta F A \sum \psi \qquad (9.46)$$

The material cost of the module $C_m [\$/\text{m}^2]$ is found with the following formula according to the mass calculation of a leg in Section 5.3:

$$C_m = (\rho_{TE} F d C_{TE} + 2\rho_s d_s C_s) N \qquad (9.47)$$

where N is the number of legs per unit area $[1/\text{m}^2]$ and C is the market price of raw materials $[\$/\text{kg}]$. Subscript s indicates substrate for both hot and cold sides.

This tool is designed to calculate a co-generation system performance with utilizing thermoelectric as the topping cycle over a steam or gas turbine power generator. If the system does not have such a bottoming cycle, the user can simply give '0' for efficiency, initial cost [$/kW], and thermal resistance [Km^2/W] of turbine. Then, the gas temperature must be the thermal ground, typically ambient air temperature. Otherwise, the gas temperature means the refrigerant temperature between the cycles.

The heat sink cost is defined by footprint area A_{HS} [m^2] of the heat flow, which is equal to the module area in most of the cases. Fuel price is found from market price [$/kg] and the calorific value, typically lower heating value (LHV) [J/kg] of the fuel.

9.5   Material Properties Simulator (BTEsolver Tool)

The third simulation tool is a tool that can be used to calculate electron transport properties such as S, σ, the power factor, and ZT for thermoelectric materials. The calculations are based on the linearized BTE within the relaxation time approximation, which is one of the most successful electron transport models for bulk semiconductors. Hence, the tool is called the BTEsolver tool [5]. The electronic band structure and scattering characteristics of the material simulated are user inputs that are needed for the calculations. The thermal transport properties such as the lattice thermal conductivity cannot be solved by this tool. A constant lattice thermal conductivity is instead provided by the user as an input value, which is then included in the total thermal conductivity to obtain the final figure of merit ZT.

The BTEsolver tool provides several analysis options that are useful for educating students on the physical concepts of differential conductivity, enhancement of Seebeck coefficient by electron energy filtering scheme, and doping optimization. Figure 9.10 shows the sequence of user inputs and selection of the analysis options and execution of simulation using the BTEsolver tool. Details about the simulation principles and analysis sub-tools are discussed in the following subsections.

Phase 1: Enter band structure information.

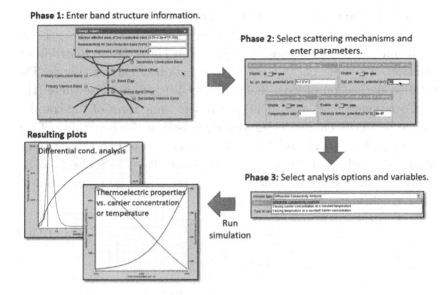

Fig. 9.10:   Simulation procedure of BTEsolver tool involving four steps.

9.5.1   *Linearized Boltzmann transport equations*

The BTE depict the electric flux and the heat flux carried by the charge carriers under external forces such as temperature gradients, and electric fields. Within the limit of weak forces, these fluxes are approximately in a linear proportion to the external forces. We deal with this linear regime only, using the so-called linearized BTE. In this model, the electric and heat fluxes can be described as linear functions of the two forces, the electric field, $\vec{E}$, and the temperature gradient, $-\nabla T$, given by [1]

$$\vec{J} = e^2 L_0 \vec{E} - \frac{e}{T} L_1(-\nabla T) \qquad (9.48)$$

$$\vec{q} = -e L_1 \vec{J} + \frac{1}{T} L_2(-\nabla T) \qquad (9.49)$$

The coefficients L_i ($i = 0, 1,$ and 2) in these equations are given by the following integral:

$$L_i = \int_0^\infty \tau(E) v^2(E) \rho_{\mathrm{DOS}}(E)(E - E_F)^i \left(-\frac{\partial f_0}{\partial E}\right) dE \qquad (9.50)$$

where E is the carrier energy, τ is the relaxation time or scattering time of carriers, v is the carrier velocity in one direction, ρ_{DOS} is the electronic density of states, and f_0 is the Fermi-Dirac distribution function. All these parameters are a function of carrier energy. The detailed derivation of Eqs. (9.48) and (9.49) is found elsewhere [1].

Equations (9.48) and (9.49) are in the same form of the coupled charge and heat current equations discussed in the beginning of the chapter in Eqs. (9.1) and (9.2). By relating these two sets of equations, one finds that the electrical conductivity, the Seebeck coefficient, and the electronic thermal conductivity can be determined by the L coefficients, respectively, as

$$\sigma = e^2 L_0 \tag{9.51}$$

$$S = -\frac{1}{eT}L_1 \tag{9.52}$$

$$\kappa_{\text{elect}} = \frac{1}{T}\left(L_2 - \frac{L_1^2}{L_0}\right) \tag{9.53}$$

Here, the integrand in the integral function of L_i shown in Eq. (9.50) except for the part $(E - E_F)^i$ is called the *differential conductivity*, σ_d, such that

$$\sigma_d = e^2 \tau(E) v^2(E) \rho_{\text{DOS}}(E)\left(-\frac{\partial f_0}{\partial E}\right) \tag{9.54}$$

because its integral over the energy range gives the total electrical conductivity by Eq. (9.51). Then, all the thermoelectric properties in Eqs. (9.51)–(9.53) can be re-written as integral functions of the differential conductivity. Considering multiple bands involving in the transport, the thermoelectric material properties are finally given as

$$\sigma = \sum_i \int \sigma_{d,i}(E)dE \tag{9.55}$$

$$S = \frac{1}{qT}\sum_i \frac{\int \sigma_{d,i}(E)(E - E_{F,i})dE}{\int \sigma_{d,i}(E)dE} \tag{9.56}$$

$$\kappa_{\text{elect}} = \frac{1}{q^2 T}\sum_i \int \sigma_{d,i}(E)(E - E_{F,i})^2 dE - S^2 \sigma T \tag{9.57}$$

where $\sigma_{d,i}$ is the differential conductivity of the i-th band of the specific type of carrier, q is $+e$ for holes, and $-e$ for electrons, and $E_{F,i}$ is the relative position of the Fermi level from the i-th band edge. The contributions from all the bands are added to obtain the transport properties for the multi-band structure.

9.5.2 Band structure, density of states, and carrier concentration

In the usual thermoelectric semiconductor materials, multiple electronic bands are involved in the carrier transport. The BTEsolver tool can accommodate a total of four bands, with maximum two conduction bands and maximum two valence bands for the material simulation, which covers most of the important TE materials. Figure 9.11 shows a screenshot of BTEsolver tool, where the user can enter band structure information for maximum four bands. In BTEsolver, the E-k dispersion relation for each band is modeled as a non-parabolic band with an effective mass m^* and a parameter called the non-parabolicity α given by

$$E(1 + \alpha E) = \frac{\hbar^2 k^2}{2m^*} \tag{9.58}$$

At low electron energies, the band can be assumed to be parabolic with $\alpha = 0$. However, for a large carrier concentration or at high

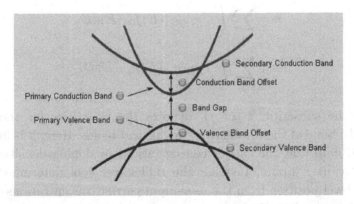

Fig. 9.11: Band structure used in BTEsolver tool with maximum of two conduction bands and two valence bands.

temperatures, it is important to use a non-parabolic band model because a large number of carriers could fill up to high energies, where deviation from the parabolic dispersion relation is severer.

The user needs to enter the single-valley effective mass, the non-parabolicity, and the band degeneracy for each band by clicking the selective button located near the band drawing in BTEsolver. In addition, the band offsets and the band gap values need to be entered. These values can be entered as a function of temperature or material composition in BTEsolver.

The density of states (DOS) for a non-parabolic band is given by

$$\rho_{\text{DOS}}(E) = D \frac{\sqrt{2}(m^*)^{\frac{3}{2}}}{\pi^2 \hbar^3} \sqrt{E + \alpha E^2}(1 + 2\alpha E) \qquad (9.59)$$

where D is the band degeneracy. The electron velocity in one direction $v(E)$ used in the differential conductivity is given by

$$v(E) = \sqrt{\frac{2}{3m^*}} \frac{\sqrt{E + \alpha E^2}}{(1 + 2\alpha E)} \qquad (9.60)$$

The electron (or hole) carrier concentration is obtained by summing the carrier concentration from each band in the conduction (or valence) bands. They are obtained based on the Fermi-Dirac distribution and the DOS in the energy space, respectively, as

$$n = \sum_i \int_0^\infty \rho_{\text{DOS},i}(E) f_{0,i}(E) dE \qquad (9.61)$$

$$p = \sum_j \int_0^\infty \rho_{\text{DOS},j}(E) f_{0,j}(E) dE \qquad (9.62)$$

where the subscript, i (or j), denotes the quantities for the i-th (or j-th) band of the conduction (or valence) bands. Here, it is important to note that the carrier concentrations or doping densities are typically user inputs. Instead, the BTEsolver tool determines the Fermi-level position from the user-input carrier concentrations using Eqs. (9.61) and (9.62) by a reverse search algorithm, which is then used to compute the transport properties.

9.5.3   *Bipolar transport*

In a typical narrow-bandgap semiconductor, there always exist two types of charge carriers involved in the carrier transport, which are electrons and holes, although in highly doped materials, it is possible that the concentration of one carrier type is much higher than that of the other type, so that only one carrier type can be assumed. In the BTEsolver tool, the thermoelectric properties are calculated for each carrier type using Eqs. (9.55)–(9.57) and then the total properties are obtained using the two quantities from the individual carrier types.

The total electrical conductivity and the total Seebeck coefficient in the bipolar transport regime are obtained as

$$\sigma = \sigma_e + \sigma_h \tag{9.63}$$

$$S = \frac{\sigma_e S_e + \sigma_h S_h}{\sigma_e + \sigma_h} \tag{9.64}$$

where the subscripts e and h denote the partial properties of electrons and holes, respectively. The total Seebeck coefficient is a weighted average of the two partial properties of electrons and holes with their electrical conductivities as the weights.

The total electronic thermal conductivity is not only the sum of the partial electronic thermal conductivities of electrons and holes, but another term from the *bipolar thermodiffusion effect* must be added. The bipolar electronic thermal conductivity is given by [11]

$$\kappa_{\mathrm{bi}} = \frac{\sigma_e \sigma_h}{\sigma_e + \sigma_h} (S_e - S_h)^2 T \tag{9.65}$$

The bipolar electronic thermal conductivity can be very large, particularly at high temperatures and at low doping levels. κ_{bi} increases with temperature almost exponentially and has a peak in the intrinsic regime. The total electronic thermal conductivity is then obtained as

$$\kappa_{\mathrm{elect}} = \kappa_{\mathrm{elect},e} + \kappa_{\mathrm{elect},h} + \kappa_{\mathrm{bi}} \tag{9.66}$$

The thermal conductivity that is experimentally measured is the sum of the lattice thermal conductivity and the electronic thermal conductivity such that

$$\kappa = \kappa_{\text{lat}} + \kappa_{\text{elect}} \tag{9.67}$$

One can extract the lattice thermal conductivity from the measured total thermal conductivity and the calculated electronic thermal conductivity using Eq. (9.67). Then, the obtained lattice thermal conductivity is assumed to be constant at a given temperature to analyze the variation of the material properties as a function of carrier concentration.

9.5.4 Scattering characteristics

In BTEsolver, several major scattering mechanisms are included to obtain the total scattering time as a function of carrier energy. One of the most significant scattering mechanisms in most of the state-of-the-art TE materials is the acoustic phonon deformation potential scattering. The energy-dependent scattering time by acoustic phonon deformation potential is given by [12]

$$\tau_{AC}(E) = \frac{\pi \hbar^4 C_l}{\sqrt{2}(m^*)^{3/2} D_a^2 k_B T} \frac{1}{\sqrt{E}} \tag{9.68}$$

where C_l is the elastic constant, and D_a is the acoustic phonon deformation potential. The optical phonon deformation potential scattering is also included in the tool, for which the same Eq. (9.68) is used except with the optical phonon deformation potential D_o replacing D_a.

Other important scattering mechanisms are the ionized impurity scattering and the polar optical phonon (POP) scattering. The scattering times by ionized impurities (II) and POPs are given, respectively, by [13, 14]

$$\tau_{II}(E) = \frac{16\pi \sqrt{2m^*} \varepsilon_0^2}{N_{II} e^4} \left[\ln \left(\frac{\delta_0 + 1}{\delta_0} \right) - \frac{\delta_0}{1 + \delta_0} \right]^{-1} E^{\frac{3}{2}} \tag{9.69}$$

$$\tau_{POP}(E) = \frac{\hbar^2}{\sqrt{2m^*} e^2 k_B T (\varepsilon_\infty^{-1} - \varepsilon_0^{-1})}$$

$$\times \left[1 - \delta_\infty \ln \left(\frac{\delta_\infty + 1}{\delta_\infty} \right) \right]^{-1} E^{\frac{1}{2}} \tag{9.70}$$

where ε_∞ and ε_0 are, respectively, the high-frequency and static permittivities, $\delta_0 = \frac{\hbar^2}{8m^* r_0^2 E}$, and $\delta_\infty = \frac{\hbar^2}{8m^* r_\infty^2 E}$. Here, r_0 is the static screening length given from

$$\frac{1}{r_0^2} = \frac{e^2}{\varepsilon_0} \int_0^\infty \left(-\frac{\partial f_0}{\partial E}\right) \rho_{\text{DOS}}(E) dE \qquad (9.71)$$

and r_∞ is the high-frequency screening length given by the same equation, but with ε_∞ replacing ε_0.

The scattering time by the alloy scattering is given by Harrison and Hauser's expression as [15]

$$\tau_{AL}(E) = \frac{8\sqrt{2}\hbar^4}{3\pi(m^*)^{3/2}\Omega U_a^2 s} \frac{1}{\sqrt{E + \alpha E^2}(1 + 2\alpha E)} \qquad (9.72)$$

where Ω is the primitive cell volume, U_a is the alloy scattering potential, and s is the randomness of the alloy that describes the effect of ordering on the scattering rate. In general, $0 \le s \le 1$, with $s = 1$ for a perfectly random alloy, and $s = 0$ for a perfectly ordered alloy. The evaluation of s is complicated, and depends on the scattering potential, U_a [15]. In BTEsolver, we set $s = 1$ assuming a perfectly random alloy.

The scattering time by the defect vacancy (non-Coulomb) deformation potential scattering is given by [16]

$$\tau_{\text{DV}}(E) = \frac{\pi\hbar^4}{\sqrt{2}(m^*)^{3/2}N_V U_V^2 s} \frac{1}{\sqrt{E + \alpha E^2}(1 + 2\alpha E)} \qquad (9.73)$$

where N_V is the non-ionized defect density, and U_V is the short-range potential of the defects.

Within the assumption that scattering events are independent of each other, the total energy-dependent scattering time can be obtained by the Matthiessen's rule such that

$$\frac{1}{\tau(E)} = \frac{1}{\tau_{AC}(E)} + \frac{1}{\tau_{II}(E)} + \frac{1}{\tau_{\text{POP}}(E)}$$

$$+ \frac{1}{\tau_{AL}(E)} + \frac{1}{\tau_{DV}(E)} + \cdots \qquad (9.74)$$

This total scattering time is plugged into Eq. (9.54) to calculate the differential conductivity as a function of energy, which is then used to calculate the thermoelectric properties using Eqs. (9.55)–(9.57).

9.5.5   *Doping optimization*

BTEsolver is a useful tool for TE material optimization with doping and temperature. It has been actively utilized in several recent research, and the results directly obtained from BTEsolver have been published in recent papers [1, 11, 17, 18]. In Phase 3, shown in Fig. 9.12, the user can select the option of 'varying carrier concentration at a constant temperature' for doping optimization at a given temperature, or select the option of 'varying temperature at a constant carrier concentration' for performance optimization over a broad temperature range for a fixed carrier concentration. One can run the tool several times with different values of fixed parameter to obtain the results over the two parameter ranges.

Figure 9.12 shows an example of simulation results for p-type PbTe by BTEsolver. Once the simulation is done, these plots are displayed as a function of the independent variable, e.g. hole concentration in this case at 600 K. As shown in Fig. 9.12(a), there is a trade-off between the Seebeck coefficient and the electrical conductivity via the carrier concentration at the high concentration region. At low concentrations, the bipolar effect takes place and the Seebeck coefficient drops significantly. BTEsolver also shows the thermal conductivity as a function of hole concentration, as shown in Fig. 9.12(b). The electronic thermal conductivity increases sharply with carrier concentration due to their Wiedemann–Franz relation at high concentrations. The bipolar thermal conductivity appears significantly at low concentrations, which dominates the electronic thermal conductivity in this regime. The total thermal conductivity is the sum of the electronic thermal conductivity and the constant lattice thermal conductivity, which is obtained from experiments, e.g. 1.2 W/(m K) in this case in the figure.

The final ZT and power factor are displayed in Fig. 9.12(c). It is found that a maximum ZT of 1.7 can be obtained from this material at 600 K with the hole concentration of 9.6×10^{19} cm^{-3}. The ZT has

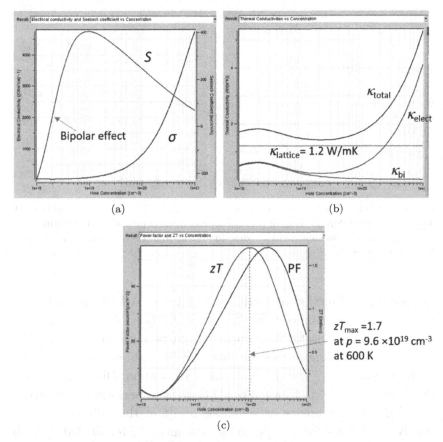

Fig. 9.12: Doping optimization of p-type PbTe by BTEsolver at 600 K. (a) Calculated Seebeck coefficient and electrical conductivity, (b) thermal conductivities, and (c) ZT and power factor as a function of hole concentration from 10^{18} to 10^{21} cm^{-3} at 600 K. In (b), a constant lattice thermal conductivity of 1.2 W/(m K) was added.

its maximum at a slightly lower concentration than the power factor has because of the increased thermal conductivity with increasing concentration.

9.5.6 *Differential conductivity analysis*

Apart from the material optimization with doping and temperature, BTEsolver provides an option where the differential conductivity is

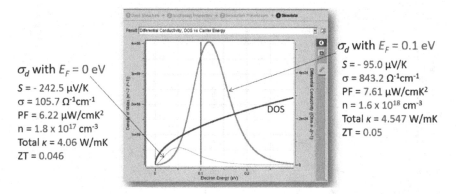

σ_d with $E_F = 0$ eV

S = - 242.5 μV/K
σ = 105.7 Ω⁻¹cm⁻¹

PF = 6.22 μW/cmK²
n = 1.8 x 10¹⁷ cm⁻³
Total κ = 4.06 W/mK
ZT = 0.046

σ_d with $E_F = 0.1$ eV

S = - 95.0 μV/K
σ = 843.2 Ω⁻¹cm⁻¹
PF = 7.61 μW/cmK²
n = 1.6 x 10¹⁸ cm⁻³
Total κ = 4.547 W/mK
ZT = 0.05

Fig. 9.13: Differential conductivity analysis by BTEsolver for n-type InGaAs at 300 K with two different Fermi-level positions of 0 and 0.1 eV above the conduction band minimum. Also shown is the variation of the material properties for the two Fermi-level positions.

displayed as a function of carrier energy with fixed Fermi-level position, doping and temperature. This visualization of differential conductivity is very useful for education, particularly by showing the effect of the differential conductivity and Fermi-level position on the Seebeck coefficient. The user can select the option 'Differential Conductivity Analysis' in Phase 3 in BTEsolver to perform the analysis.

Figure 9.13 presents an example result of the differential conductivity analysis for n-type InGaAs semiconductor at 300 K. Two Fermi-level positions were selected, 0 and 0.1 eV above the conduction band minimum, and the differential conductivity σ_d was plotted as a function of electron energy for each Fermi-level position. The DOS of this material is also plotted as a curve indicated as DOS in the figure and the vertical lines are the Fermi-level positions. As shown in the figure, when the Fermi level is increased from 0 to 0.1 eV, the lower energy portion of differential conductivity below the Fermi level is enlarged, which reduces the asymmetry in the differential conductivity portions below and above the Fermi level. This is the key reason for the reduction in Seebeck coefficient.

BTEsolver calculated the resulting material properties for each Fermi-level position. As shown in the figure, the magnitude of Seebeck coefficient decreased substantially from 242.5 to 95 μV/K when the Fermi-level position increased from 0 to 0.1 eV. At the same time, the electrical conductivity increased from 105.7 to 843.2 $\Omega^{-1}\,\mathrm{cm}^{-1}$ due to the increased Fermi level or the corresponding increase in the carrier concentration. Due to this trade-off between the Seebeck coefficient and the electrical conductivity via the Fermi level or doping concentration, along with the change in the electronic thermal conductivity, the ZT does not change much by this Fermi-level change.

9.5.7 *Electron energy filtering analysis*

In addition to the differential conductivity analysis, BTEsolver can add the electron energy filtering effect in the analysis. This is useful to educate students about the energy filtering effect in thermoelectrics. In Phase 3, the user can add a non-zero cut-off energy level, below which all the carriers are prevented from contributing to the transport. This is a perfect filtering case that can be simulated in BTEsolver.

Figure 9.14 shows an example result of the differential conductivity analysis with the energy filtering effect. The cut-off energy

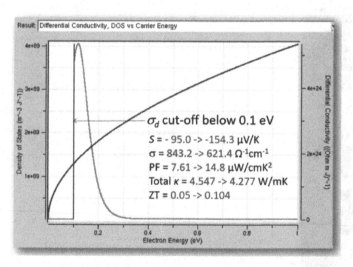

Fig. 9.14: Impact of electron energy filtering simulated using BTEsolver with a cut-off energy at 0.1 eV. Also shown is the variation of material properties before and after the filtering effect.

was set at 0.1 eV. As a result, the differential conductivity below the cut-off energy was completely removed to simulate the perfect filtering effect. Because of this effect, the material properties are changed significantly. As shown in Fig. 9.13, the magnitude of Seebeck coefficient increases dramatically from 95.0 to 154.3 μV/K due to the filtering effect. The electrical conductivity drops at the same time, but the net power factor increased by almost a factor of two due to the large Seebeck enhancement. Consequently, ZT increased from 0.05 to 0.104 by the same factor.

9.6   Online Course

An online course entitled 'Thermoelectricity: from Atoms to Systems' has been developed on the nanoHUB-U (Fig. 9.15) platform

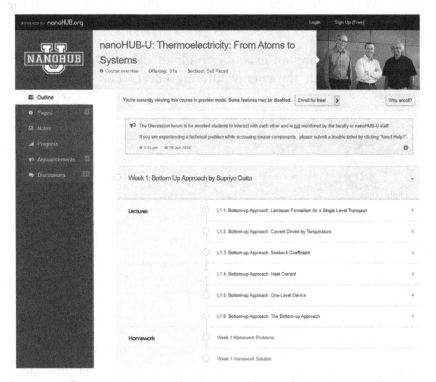

Fig. 9.15:   Front page of the online course 'Thermoelectricity: From Atoms to Systems' on nanoHUB.org.

for educating the public about the fundamental principles of thermoelectricity and the state-of-the-art technologies at multi-scale [6]. The course can be accessed any time at https://nanohub. org/courses/TEAS. When it was first-time offered in summer 2016, this course had a dual delivery format, i.e. instructor-led and self-paced modes. The instructor-led format was offered one time and the learners should complete the course in a 5-week time frame. During the format, learners can interact with the instructors nearly real-time on the online discussion platform offered by nanoHUB-U, and receive optional feedback on homeworks and exams. Since the first period, the course is currently offered in the self-paced mode only.

In this online course, learners watch 5–7 pre-recorded lecture videos per week for the 5-week period at their own paces, solve exercise problems as homework and take an optional exam after each week to verify their learning. Three instructors participated in the course. The course content is summarized in Table 9.1.

In Weeks 4 and 5, the simulation tools are used to solve the homework problems. The THERMO tool is used to simulate and optimize thin-film thermoelectric cooler and power generator devices as well as double-segmented elements in Week 4. The TEDEV tool is also used in that week to simulate the cost-performance trade-off in a thermoelectric system. The BTEsolver tool is practiced in Week 5 to simulate PbTe and Mg_2SnSi materials as examples.

During the instructor-led period, several survey questionnaires were asked to learners to investigate the impacts of using the simulation tools on learning. Learning gains, instructional support effect, and student perceptions about the course were also investigated. According to the survey, learners significantly increased their understanding of thermoelectric devices concepts by utilizing the simulation tools. More details about the survey results are found elsewhere [19].

9.7 Summary

Several online simulation tools have been developed for education and research in the field of thermoelectrics. Built upon the nanoHUB

Table 9.1: Contents of the online course 'Thermoelectricity: From Atoms to Systems'.

Period (Lecturer)	Subjects
Week 1 (Supriyo Datta)	Title: Bottom-up Approach
	L1.1: Landauer Formalism for a Single-Level Transport
	L1.2: Current Driven by Temperature
	L1.3: Seebeck Coefficient
	L1.4: Heat Current
Week 2 (Mark Lundstrom)	Title: Thermoelectric Transport Parameters
	L2.0: Short Introduction
	L2.1: Landauer-Boltzmann Approach
	L2.2: Thermoelectric Transport Coefficients
	L2.3: Devices and Materials
	L2.4: Novel Materials and Structures
	L2.5: Lattice Thermal Conductivity
	L2.6 (Bonus): Boltzmann Transport Equation
	L2.7 (Bonus): Using Full Band Dispersions (ex. of Bi_2Te_3)
Week 3 (Ali Shakouri)	Title: Nanoscale and Macroscale Characterization
	L3.0: Introduction and Motivation
	L3.1: Micro/Nano Scale Temperature Measurement (Part 1)
	L3.2: Micro/Nano Scale Temperature Measurement (Part 2)
	L3.3: Thermoreflectance, Micro-Raman, Suspended Heaters
	L3.4: Thin film Thermoelectric Characterization
	L3.5: Thermoreflectance Laser Characterization
	L3.6: Overview of Week 3
Week 4 (Ali Shakouri)	Title: Thermoelectric Systems
	L4.1: Thermoelectric Cooling and Power Generation Applications
	L4.2: Thermoelectric Cost/Efficiency Trade Off
	L4.3: Microrefrigerator on a Chip
	L4.4: Graded Materials, TE Leg Geometry Impact
	L4.5: Ballistic Thermionic Coolers and Nonlinear Peltier
	L4.6: Overview of Week 4

(*Continued*)

Table 9.1: (*Continued*)

Period (Lecturer)	Subjects
Week 5 (Ali Shakouri)	Title: Recent Advances in Thermoelectric Materials and Physics L5.1: Thermionics vs. Thermoelectrics L5.2: Semiconductors with Embedded Nanoparticles L5.3: State-of-the-art Thermoelectric Materials, Optimum Bandgap L5.4: Skutterudites, Oxide thermoelectrics, Spin Seebeck, Resonant State L5.5: Ideal Thermoelectrics, Carnot vs. Curzon-Ahlborn Limits, Some Open Questions L5.6: Overview of Week 5, Recent Reviews

platform, they offer easy graphical user interfaces and capabilities of simulating thermoelectric devices and systems as a function of various design parameters for convenient design optimization. The THERMO tool provides simulation on thin-film thermoelectric devices in both cooling and power generation applications. The ADVTE tool provides realistic performance prediction of multi-element thermoelectric modules by taking into account temperature-dependent material properties for large heat-input applications such as high temperature waste heat recovery. The TEDEV tool offers analysis methods for cost-performance trade-off in thermoelectric systems. The BTEsolver tool is capable of calculating the thermoelectric material properties based on the electronic band structure and scattering characteristics of the target material. Additionally, an online course has been developed on the same platform in which these simulation tools are utilized to enhance the learning of the state-of-the-art thermoelectric technologies.

References

[1] J.-H. Bahk and A. Shakouri, Electron transport engineering by nanostructures for efficient thermoelectrics. In *Nanoscale Thermoelectrics*, X. Wang and Z. Wang, (Eds.), Springer (2013).

[2] J.-H. Bahk, M. Youngs, Z. Schaffter, K. Yazawa and A. Shakouri, Thin Film and Multi-element Thermoelectric Devices simulator, https://nanohub.org/resources/thermo (DOI: 10.4231/D33T9D71B), 2014.

[3] J.-H. Bahk, K. Margatan, K. Yazawa and A. Shakouri, Advanced Thermoelectric Power Generation Simulator for Waste Heat Recovery and Energy Harvesting, https://nanohub.org/resources/advte (DOI: 10.4231/D3XW47X4X), 2016.

[4] K. Yazawa, K. Margatan, J.-H. Bahk and A. Shakouri, Thermoelectric system performance optimization and cost analysis, https://nanohub.org/resources/tedev (DOI: 10.4231/D3028PD8C), 2014.

[5] J.-H. Bahk, R. B. Prost, K. Margatan and A.Shakouri, Linearized Boltzmann Transport Calculator for Thermoelectric Materials, https://nanohub.org/resources/btesolver (DOI: 10.4231/D37M04151), 2014.

[6] A. Shakouri, M. Lundstrom and S. Datta. [Online]. Available: https://nanohub.org/courses/TEAS.

[7] D. Vashaee, J. Christofferson, Y. Zhang, A. Shakouri, G. Zeng, C. LaBounty, X. Fan, J. Piprek, J. Bowers and E. Croke, Modeling and optimization of single-element bulk SiGe thin-film coolers, *Nanosc. Microsc. Therm.*, **9**, 99–118 (2005).

[8] S. Lee, S. Song, V. Au and K. P. Moran, Constriction/spreading resistance model for electronics packaging, in *ASME/JSME Thermal Engineering Conference* (1995).

[9] Z. Bian and A. Shakouri, Beating the maximum cooling limit with graded thermoelectric materials, *Appl. Phys. Lett.*, **89**, 212101 (2006).

[10] G. Nolas, J. Sharp and J. Goldsmid, *Thermoelectrics: Basic Principles and New Materials Developments*, Springer (2001).

[11] J.-H. Bahk and A. Shakouri, Minority carrier blocking to enhance the thermoelectric figure of merit in narrow band gap semiconductors, *Phys. Rev. B*, **93**, 165209 (2016).

[12] D. I. Bilc, S. D. Mahanti and M. G. Kanatzidis, Electronic transport properties of PbTe and AgPbmSbTe2+m systems, *Phys. Rev. B*, **74**, 125202 (2006).

[13] Z. Bian, M. Zebarjadi, R. Singh, Y. Ezzahri, A. Shakouri, G. Zeng, J.-H. Bahk, J. E. Bowers, J. M. O. Zide and A. C. Gossard, Cross-plane Seebeck coefficient and Lorenz number in superlattices, *Phys. Rev. B*, **76**, 205311 (2007).

[14] C. J. Vineis, T. C. Harman, S. D. Calawa, M. P. Walsh, R. E. Reeder, R. Singh and A. Shakouri, Carrier concentration and temperature dependence of the electronic transport properties of epitaxial PbTe and PbTe/PbSe nanodot superlattices, *Phys. Rev. B*, **77**, 235202 (2008).

[15] J. W. Harrison and J. R. Hauser, Alloy scattering in ternary III-V compounds, *Phys. Rev. B*, **13**, 5347 (1976).

[16] D. M. Zayachuk, The dominant mechanisms of charge-carrier scattering in lead telluride, *Semiconductors*, **31**, 173–176 (1997).

[17] J.-H. Bahk, Z. Bian and A. Shakouri, Electron transport modeling and energy filtering for efficient thermoelectric Mg2Si1-xSnx solid solutions, *Phys. Rev. B*, **89**, 075204 (2014).

[18] J.-H. Bahk and A. Shakouri, Enhancing the thermoelectric figure of merit through the reduction of bipolar thermal conductivity with heterostructure barriers, *Appl. Phys. Lett.*, **105**, 052106 (2014).

[19] M. d. R. Uribe, A. Magana, J.-H. Bahk and A. Shakouri, Computational simulations as virtual laboratories for online engineering education: A case study in the field of thermoelectricity, *Comp. Appl. Eng. Educ.*, **24**, 428–442 (2016).

Chapter 10

Future of Technology

10.1 Energiology

Energiology is a coined word representing the influence and impact of energy in all activities of humanity. This section discusses the potential future in energy, environment, and the quality of life, including what if thermoelectric were used more broadly. As we talk about the impact of energy on civilization, one cannot underestimate the huge impact of concentrated and transportable fossil sources (coal, petroleum, and gas) and how they have enabled unprecedented comfort and major industries in the last 200 years. French energy analyst, Jean Michel Jancovici, makes an interesting analogy with human and animal workforce, which were the basis of advancement in human society for thousands of years before the fossil fuel era. Jancovici uses the term 'energy slave' to highlight what has really enabled the western lifestyle which is envy of most countries on Earth. Figure 10.1 shows the equivalent power of 1 l of petroleum compared to human forces. An interesting side note is that since the basis of today's economic prosperity is the growth in gross domestic product (GDP) and economic theories are not well equipped to deal with finite resources (environment, fossil fuels), one has to really develop new measures of development if we are serious in reducing our dependence on finite resources and want to keep beautiful resources on Earth for our children and many generations to come. As Thomas Jefferson noted in 1789, 'Then I say the earth belongs to each...generation during its course, fully and in its own right. The second generation receives

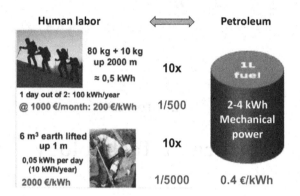

Fig. 10.1:   Energy content of human labor for hiking or for Earth excavation and its cost based on minimum salary compared with energy and price of 1 L of petroleum according to Jancovici [1].

it clear of the debts and encumbrances, the third of the second, and so on. For if the first could charge it with a debt, then the earth would belong to the dead and not to the living generation. Then, no generation can contract debts greater than may be paid during the course of its own existence.'

In the 1990s, pure electrical 'Plug-in' vehicles made a debut as GM [2] and Honda [3] introduced the passenger cars to the market. This was followed by Chrysler [4] and Ford [5] in commercial segment and others. The attention was immediately raised when the State of California set the bar to zero emission vehicles as the ultimate environmental goal at the time. Interestingly, the first era of fully electrical drive vehicles was more than a century before it. Following the invention of lead-acid batteries in 1859, several developments happened in 1890s. It is believed that 40% of the U.S. registered vehicles were electrically driven by 1900. The big difference in their usage in the society as compared to gasoline engine was 'battery charge prior to use,' regardless of the kind of vehicle — passenger car, commercial or any other motive equipment. This is also true even for hybrid or fuel-based electrical vehicles. The hype of electric vehicles in early 1990s was replaced by hydrogen, biofuels, plug-in hybrid, etc. in the subsequent years. As highlighted in a nice article [6] in Nature Energy, the phenomenon of 'energy *du jour*' is wasteful as we put a lot of resources for a short-term miracle, and when we don't achieve the ambitious goals, we turn to a different source of energy. Identifying new, reliable, low cost and environmentally responsible energy sources requires long-term investment and major changes in energy distribution and usage infrastructure. We need consistent planning and critical analysis of options that are decoupled from short-term political changes that are typical in democratic countries.

Current electrical power infrastructure of almost all countries is more or less the same as a century ago, as shown in Fig. 10.2. Centralized large power plants are located in designated areas mainly independent of their primary energy resources. Majority of the power generation still highly depends on fossil fuels (coal, petroleum and natural gas). Even those who use nuclear fission or natural resources, such as solar, wind, geothermal, or hydro, are likely be centralized facilities, except the residential scale roof-top solar photovoltaic (PV) cells. These power sources are connected to a power grid that is physically networked over long distances. Through the transformer, the power is finally distributed to the house or building. Does this infrastructure really match the future large potential demand? Would it be on-demand in any place? In order to find the solution, the following data will be useful to revisit.

Figure 10.3 shows a summary of global energy usage flow adapted from Cullen and Allwood [7]. Energy resources are shown on the left and they are converted to products/services via conversion

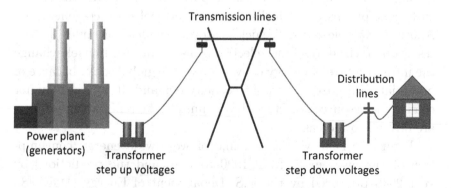

Fig. 10.2: Diagram of a typical electrical power infrastructure.

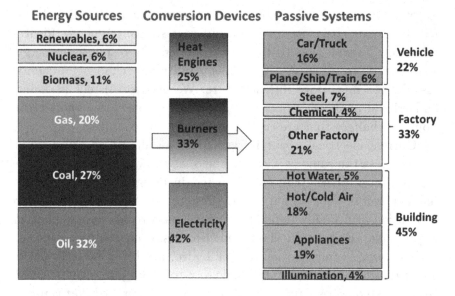

Fig. 10.3:   Global map of energy flow from resource to service.
For the source of original data, see Ref. [7].

methods/devices (heat engines, furnaces/burners and electrical power plants). Almost half of energy usage is for buildings (appliances, hot/cold air, hot water and illumination). Buildings are distributed. Burning a high exergy fuel (natural gas or electricity) to produce low quality heat (20–80°C) for air or water is a huge waste of the ability to do work. There is an opportunity to increase the exergy with heat pumps or with distributed thermoelectric co-generators. Another example is for 'Vehicle' category, which consumes 22% of resources. Hybrid fossil fuel/electric transportation will likely change the infrastructure of energy conversion at the global scale, because of potential abrupt expansion of electricity demand. It may or may not change the resource structure, but definitely the renewable resources are desired to increase.

Figure 10.4 shows the timeline of worldwide energy consumption of various sources from 1990 to 2010 and the predictions by year 2040 produced by the U.S. Department of Energy (DoE) [8]. The energy unit in the y-axis is Quads, where 1 Quad is equal to

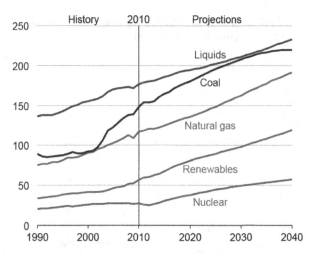

Fig. 10.4: Worldwide energy consumption in history and projection, published in year 2013 based on the 2010 data. It is advisable to compare the 2020 data in the two graphs with Fig. 10.5.

1.055×10^{18} Joules or 0.293×10^{12} kWh. According to the predictions, while renewables double in about twenty years, the usage of fossil resources will also keep increasing. Coal will be further consumed, more than the previous prediction released in 2000, even though shale gas has been developed quite well between 2008 and 2020 in the continental U.S. This will have significant environmental consequences since many medium or small range power plants in rapidly developing countries such as China and India are running on coal and they have the potential to satisfy the large increase in electricity demands in the next few decades.

Figure 10.5 shows similar statistics by energy sectors and by resources in 2018 with prediction up to 2050, published in 2019 [9]. Both graphs are adapted from the U.S. Energy Administration.

One may notice that predictions 8 years apart can be off by up to 200%. This should be a lesson for humility. We can't really predict the future even with our most advanced computer models and data analytics. On the other hand, scenario analysis is quite useful as it shows the implications of putting emphasis on existing technology as we move forward [10].

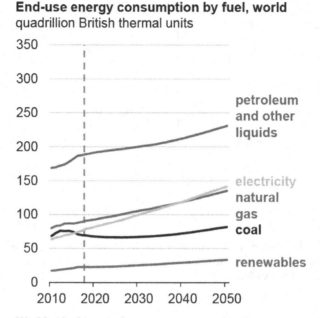

Fig. 10.5: Worldwide historical energy consumption by sectors and by fuel resources and the prediction, see Ref. [9].

The key issues identified, based on the historical data and predictions, are as follows:

1. Fossil fuel consumption will likely increase.
2. Renewable energy consumption is actually not increased as much as needed and predicted.

In the coming decades, with the rise of electrification of transport, the following challenges need to be addressed:

1. Minimizing all the losses during conversion from primary energy to energy in service (electricity).
2. Use of natural gas instead of coal without changing the current power plants.
3. Increased use of local and distributed natural and renewable energy resources.

One of the potential solutions would be the microgrid concept [11], which has been already discussed in specific applications. Distributed power generation (both electricity and natural gas) with some energy storage can support a small community. Co-generation of heat and electricity can achieve high efficiencies. However, the cost is currently too high. One of the potential research-and-development areas for thermoelectrics is for small-scale and site-specific refrigeration/heating and for co-generation together with electrical power plants.

10.2 Distributed Power Generation and Scalability

10.2.1 *Scaling to smaller dimensions*

How much energy is lost from primary sources to the actual service? The following calculation describes the primary energy efficiency of typical electric and gasoline engine vehicles.

The electrical energy produced from combustion of oil, gas, and coal resources is 54 Exa J worldwide. This is a large fraction of 189 Exa J in global energy demand (data of 2005). This means a practical energy conversion efficiency of 29%. It is only about electricity generation and of course useful heat is also co-generated but not used for electric machines. Then, the electricity loss from the power plant to the individual wall AC plugs can be more or less 10%, including transmission, distribution loss factors. The loss in charging battery caused by internal electrical resistance and across the power electronics unit is in the range 12–36% [12]. This result is based on the experimental data with non-quick charger (up to 70 amperes). Near future high power quick chargers can have additional losses. Assuming 30% loss in charging battery from the grid and 12% loss in the electrical-mechanical energy conversion in motor, grid-to-wheel energy efficiency is about 62% while DoE described it as 59–62% [13]. Hence, the primary energy efficiency of a plug-in EV is projected as 17–18%. This is equivalent or even lower than the efficiency of conventional gasoline engine cars, which is '17–21% of the energy stored in gasoline to power at the wheels'. An important takeaway here is that the new electric car technology is not currently very different in terms of consumption of primary fossil energy as long as renewable power production contributes to a small fraction of electricity generation.

In principle, gasoline engine can be a distributed miniature power plant. Hence, a hybrid car, which has both electric motors and a heat engine, can provide power to local grids especially in case of hazard or emergency conditions. Unfortunately, however, heat engines are inefficient when they are scaled to smaller dimensions. Thus, the distributed system of car engines does not scale as expected. Majority of heat engines use the volume change of gas by thermal expansion, whereas the heat loss through the wall of the controlled volume increases as 'surface-area-to-volume ratio' becomes larger by shrinking the characteristic length. In addition, manufacturing tolerance does not match up well as dimensions goes smaller.

Thermoelectric devices have unique scaling characteristics among the heat engines. Thermoelectric devices are solid-state, so they don't suffer from friction losses at small scales. Also, TE modules are built from legs of millimeters in cross-section. They are put electrically in series and thermally in parallel to generate appropriate currents with reasonable voltages. The potential of thermoelectrics as small-scale energy conversion devices has been recognized [14]. As discussed in Chapter 5, Power Generation, thermoelectric is also scalable as far as proper materials are available or developed. Another advantage of thermoelectric power generators is the potential high power-weight ratio, which is dominantly controlled by the heat sink design. The mass of thermoelectric material would be negligibly small relative to that of heat sinks. Therefore, there is a large potential to implement thermoelectric, especially for transportation vehicles regardless of terrain or aerospace even for passenger or unmanned vehicles.

The scalability of thermoelectric system is realized by a modular design by arraying unit modules. If the module generates 1 W per 1 cm × 1 cm unit, a 100 m^2 area of array built in a 1 m^3 of layered plate-frame style heat exchanger theoretically generates 1×10^6 (one mega) W as far as the hot and cold gas temperatures are uniform and adequate blowers/pumps are available externally. Note that power

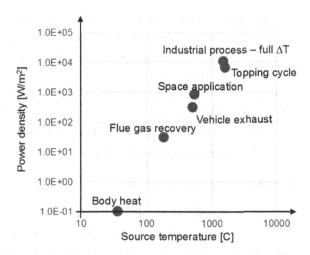

Fig. 10.6: Scalability diagram. Harvesting electrical power vs heat source temperature. The scale of power density ranges 6 figures, while solar PVs generate in the range of a few 100s W/m² depending on conditions and locations.

generation is a quadratic function of the temperature difference, hence the output power can be very different depending on the application. In Chapter 5, power harvesting trade-off by forcing the fluid through the heat exchanger is analyzed.

Figure 10.6 shows some data from the authors' analyses of thermoelectric power generation at different scales. Thermoelectric generator built in a wristwatch [15, 16] potentially generates 10s–100s of nano-Watts calculated from the available literature with temperature difference about 1°C at a source temperature of 35°C, a bit lower than that of deep inside the human body. On the higher side, topping cycle thermoelectric generators [17] for steam turbine in a large-scale (0.5 giga-W class) coal fired power plant were analyzed. Around 1400°C of hot gas temperature is available in the boiler chamber, while only 550–600°C temperature is the maximum steam for maintaining a long-term mechanical reliability of the turbine blades. The results show that the ideal thermoelectric design could generate power per unit area 6 orders of magnitude higher than body heat harvesting. Both designs are based on exactly the same principle and model, assuming similar material figures of merit $ZT_{ave} \sim 0.5$–1.

Of course, the stability of thermoelectric material, metal contacts and thermal interface layers at high temperatures need to be established. In space applications, high temperature SiGe and SbTe modules have demonstrated a long life of more than 30 years in Voyager and other spacecrafts. Terrestrial applications require designs for low cost (higher power density to reduce the use of expensive TE material) and for robust thermal cycling, which are important areas of research.

10.2.2   *Off-grid power generation*

A large population in the world is without access to any grid power. Based on the Our World in Data [18], 87% of the world's total population (6.5 Billion, year 2016) has access to electricity, which means that nearly 1 billion people don't have access. These are mostly in the least developed countries (LDC), as defined by the United Nation Development Program [19]. Available forms of energy are solar, biomass, and some types of natural gas for converting to electricity in small scale in small villages or households. Solar photovoltaics are popular, but due to the nature of their small energy density, the level of power supply is limited. Also, they limit the usable time range in daylight. The use of battery storage is very expensive in such communities.

There is potential to develop a scalable and affordable power generator based on thermoelectrics. Thermoelectric power generators in a portable cooking stove or a boiling pot already exist in the market. Some of them have cell phone charging ports, which is based on the Universal Serial Bus (USB) interface including 5V DC power line. Such equipment can provide electrical power while cooking with heat or heating the space not only for charging phone but for small electronics, such as radio and LED light, when the family gathers during dinner time.

As discussed in 10.2.1, thermoelectric is linearly scalable with the size for power output. By scaling to smaller dimensions, the power generator still retains the same level of efficiency.

Champier *et al.* discussed the danger of toxic gas from small cooking stoves especially in developing countries [21]. According to the

World Health Organization (WHO), 'Each year, close to 4 million people die prematurely from illness attributable to household air pollution from inefficient cooking practices using polluting stoves' [22]. A part of the power output from a thermoelectric generator can be used to drive fan(s) to force airflow through the stove improving the burning of wood or charcoal much better than passive natural convection. Forced air through the heat sink will also improve the power output from the thermoelectric module. As a result, safety can be improved while the efficiency of burning is increased as well and electric power is generated simultaneously.

10.2.3 *Hybrid renewables and heat pumps for dispatchability*

Power generation from renewable resources is high in demand in off-grid areas. Practically, such utilization is highly individualistic and distributed. So, trading electricity without a central grid is not easily implemented. The challenge of using natural resources is the fluctuation of electricity generation. Power output depends on the weather in addition to the limited time of daylight for solar. The demand–supply match is very hard in such isolated utilization.

We have proposed the concept of 'thermal battery' with the use of a heat pump and the hot and cold-water storages [23]. Schematic diagram is shown in Fig. 10.7. Using the fluctuated electrical power from renewables on the site, the heat pump creates the temperature difference between the two reserves and store the energy for later use.

In order to create the temperature difference between heater and chiller, the heat pump only displaces the thermal energy to create a desired temperature from excess power. Hence, the efficiency is measured by coefficient-of-performance (COP), where COP $= Q/W$ with Q being the total pumped heat and W being the electrical power input. The work focused on trans-critical CO_2 cycle since it is very efficient, where super critical phase condensation happens at the temperature of boiling water for space heating or hot water supply (40–55°C) and the evaporator temperature is easily matched to the temperature for refrigerator (10°C). In a small-scale application with

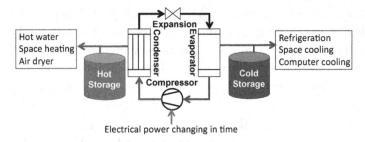

Fig. 10.7:   Thermal Battery concept diagram. The system stores the useful thermal energy converted from unstable electrical power such as wind or solar generators and supplies hot and cold heat energy simultaneously, see Ref. [23].

1 kW peak power input, the COP was found to be about 7, with individual COPs roughly 3 and 4 for chiller and boiler, respectively [24].

Thermoelectric modules can also be used as a heat pump and could replace the CO_2 compressor and expansion valve. The efficiency of thermoelectric heat pump is modest. With off-the-shelf modules with $ZT < 1$, the total COP was found to be 1.6 based on the experiments without optimization of the TE module [25]. However, there are few advantages as follows:

(1) Compact/flat system to fit in a spare space
(2) Fast response due to the small thermal mass (inertia)
(3) Cost effective and scalability for smaller system
(4) Micro miniature system

Another method of integration with intermittent renewable(s) is to use thermoelectric as a fast response generator. Especially with burning natural gas, the power output would respond quickly to the gas flow rate in the burner. Fuel consumption can be minimized to match the load all the time. This kind of supply and demand balance is hard to stabilize in a timely manner using conventional large-scale power generators. Small thermal mass or inertia of the thermoelectric system becomes beneficial.

10.3   Engineering the *ZT*

This section describes some of the latest research in thermoelectric material's figure of merit *ZT*. This has been the subject of recent reviews [26–31]. Our goal is to primarily illustrate the future potential of the materials and how they may impact the strategy and roadmap in thermoelectric energy technology landscape.

10.3.1 *Figure of merit*

A key requirement to improve the energy conversion efficiency is to increase the Seebeck coefficient (S) and the electrical conductivity (σ) while reducing the electronic and lattice contributions to thermal conductivity $(\kappa_e + \kappa_L)$. Some new physical concepts and nanostructures make it possible to modify the trade-offs between these bulk material properties through changes in the density of states, scattering rates, and interface effects on electron and phonon transport.

Detailed review is given in [26] of recent experimental and theoretical results on nanostructured materials of various dimensions: superlattices, nanowires, nanodots, and solid-state thermionic power generation devices. Most of the recent successes have been in the reduction of lattice thermal conductivity with the concurrent maintenance of good electrical conductivity and Seebeck coefficient. Several theoretical and experimental results to improve the thermoelectric power factor $(S^2\sigma)$ and to reduce the Lorenz number (σ/κ_e) are presented (Fig. 10.8).

Although the material thermoelectric figure of merit $Z[=S^2\sigma/(\kappa_e + \kappa_L)]$ is a key parameter to optimize, one has to consider the whole system in an energy conversion application.

Readers are encouraged to consult recent outstanding review articles by Snyder *et al.* [27] and Kanatzidis and colleagues [28] that discuss bulk materials. Vineis *et al.*'s [29] review focuses on nanostructured thermoelectrics and discusses some of the misconceptions in the field. Tan *et al.* [30] and He and Tritt [31] provide more recent reviews of thermoelectric materials with discussion on the physics behind the enhanced ZTs.

10.3.2 *Thermoelectric energy conversion*

The best ZT materials are found in doped semiconductors (Fig. 10.9), e.g. Ref. [32]. Insulators have poor electrical conductivity σ. Metals have relatively low Seebeck coefficients. In addition, the

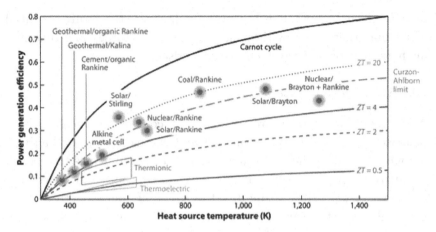

Fig. 10.8: Thermoelectric power generation efficiency versus $T_{\rm hot}$ ($T_{\rm cold} = 300\,{\rm K}$). Efficiency for conventional mechanical engines as well as the Carnot limit and the Curzon-Ahlborn limit are also shown.

Source: Reproduced with permission from *Annu. Rev. Mater. Res.*, **41**, 399–431 (2011).

thermal conductivity κ of a metal, which is dominated by electrons, is proportional to the electrical conductivity, as dictated by the Wiedemann-Franz law. The ratio of the electrical conductivity over electronic thermal conductivity (the Lorenz number σ/κ_e) is a function of the band structure, and it can be modified if, e.g. the bandwidth of the conduction band is reduced [33, 34]. It is thus hard to attain high ZT values in conventional metals.

Thermionic current [35] and hot-electron filtering [36] can improve the thermoelectric properties of degenerate semiconductors and metals, but current thin-film techniques are mostly useful for site-specific cooling of electronic devices and hot spot cooling in integrated circuits [37]. In semiconductors, the thermal conductivity consists of contributions from both electrons (κ_e) and phonons (κ_p); the majority of contribution comes from phonons. The phonon thermal conductivity can be reduced without causing too much reduction of the electrical conductivity with, e.g. alloying [38]. The traditional cooling materials are alloys of Bi_2Te_3 with Sb_2Te_3 (such as $Bi_{0.5}Sb_{1.5}Te_3$; p-type) and of Bi_2Te_3 with Bi_2Se_3 (such as $Bi_2Te_{2.7}Se_{0.3}$; n-type), with a $ZT \sim 1$ at room temperature [39].

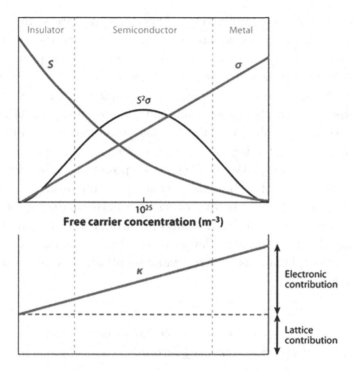

Fig. 10.9: Trade-off between electrical conductivity (σ), Seebeck coefficient (S), and thermal conductivity (κ) that involves increasing the number of free carriers from insulators to metals.

A typical power generation material is an alloy of PbTe (or SiGe), with a $ZT \sim 1$ (or 0.6) at 500°C (or 700°C).

Various means of producing ultrathin crystalline layers using metalorganic chemical vapor deposition (MOCVD) and molecular beam epitaxy (MBE) have been used to alter the bulk characteristics of materials drastically. Commercial applications include laser diodes in compact disk systems, optoelectronic components for fiber optics communication and high performance III-V transistors for RF and 5G communication devices. Drastic changes are produced by changing the crystal periodicity (by, e.g. depositing alternating layers of different crystals) or the electron dimensionality by confining the carriers in a plane (e.g. a quantum well) or in a line (e.g. a quantum wire). Quantum-effect electronic and optoelectronic devices are used in everyday applications. Thermoelectric properties of low-dimensional materials have attracted attention since the 1990s [40]. Nanostructures improve both electron and phonon transport through the use of quantum and classical size/interface effects. Researchers have explored several directions, such as quantum size effects for electrons [41, 42], thermionic emission at interfaces [43, 44], interface scattering of phonons [45, 46], and the engineering of energy dependency of scattering times using embedded nanoparticles [47]. There are new approaches to enhance the power factor in such materials as well. Recently, Heremans *et al.* [48] enhanced the Seebeck coefficient of bulk PbTe by distorting the electronic density of states and by engineering the band structure through the introduction of resonant thallium impurity levels in the bulk material. There are alternative explanations that put more emphasis on the role of multiple valleys [49].

10.3.3   *Trade-off between electrical conductivity and the Seebeck coefficient*

The trade-off between electrical conductivity and the Seebeck coefficient may be puzzling but it can be explained if we consider the concept of differential conductivity [50–52]. The analysis of differential conductivity can be performed using one of our online simulation tools at nanoHUB.org. [53], see Section 9.5.6 for more information. When the Fermi energy is close to the band edge, the density of states is asymmetric with respect to the Fermi level. This means that, e.g. for an n-type material, more states are available for transport above the Fermi energy than below it. As we increase the doping in the material, the Fermi energy moves deeper in the band, and the differential conductivity becomes more symmetric with respect to the Fermi energy. This is because the density of states has a square root dependency on energy for any typical 3D single-band crystal. Thus, using lower-dimensional semiconductors can inherently improve the thermoelectric power factor $S^2\sigma$ by creating sharp features in the electronic density of state [26]. Another possibility is to use an appropriate hot-electron filter (potential barrier) that selectively scatters cold electrons. Here, in the near-linear transport regime, hot electrons denote carriers that contribute to electrical conduction with energies higher than the Fermi level, and cold electrons have energies lower than the Fermi level. This nomenclature is different in device physics, in which hot carriers are typically non-equilibrium populations that can be built up under high electric fields. The denominator of Eq. (10.1), the formulation of Seebeck coefficient, is proportional to electrical conductivity. As the doping in the material and the conductivity increase, we need larger asymmetry in density of states if we want to keep the Seebeck coefficient high.

$$S \equiv \frac{k_B}{q} \frac{\int \sigma\left(E\right) \frac{(E-E_F)}{k_B T} \left(-\frac{\partial f_{eq}}{\partial E}\right) dE}{\int \sigma\left(E\right) \left(-\frac{\partial f_{eq}}{\partial E}\right) dE} \propto \langle E - E_F \rangle \qquad (10.1)$$

Here, q is the unit charge, k_B is Boltzmann constant, E_F is Fermi energy, and $\sigma(E)$ is a measure of the contribution of electrons with energy E to the total conductivity, and f_{eq} is the Fermi-Dirac distribution function. The trade-off between electrical conductivity and the Seebeck coefficient also originates from the fundamental trade-off between electronic density of states and electron group velocity (given by electron effective mass) in crystalline solids [54]. This is manifested by the fact that solids with a high electron effective mass and/or multiple valleys have large densities of states but lower mobilities. The shape of the density of states often dominates the overall performance, and materials with heavy electron masses and multiple valleys have good potential for high ZT. The requirement for a high effective mass is not universal; good thermoelectric materials can be found within a wide range of effective masses and mobilities from low-mobility, high-effective-mass polaron conductors (e.g. oxides, chalcogenides) to high-mobility, low-effective-mass semiconductors (e.g. SiGe). High mobility and low effective mass are typically found in materials made from elements with small electronegativity differences, whereas high effective masses and low mobilities are found in materials with narrow bands such as ionic compounds. Because the density of states is related to the whole dispersion relation in the momentum space, whereas the electron group velocity is related to the curvature of the band in a given direction, there should be potential opportunities to optimize an ideal anisotropic thermoelectric material. Electron effective mass should be small in the direction of transport, whereas there are many states (heavy mass) in the transverse direction.

10.3.4 *Optimum band gap of thermoelectric material*

Sofo and Mahan, in their very nice paper [55], generalized the work of Chasmar and Stratton [56] and derived the optimum band gap of thermoelectric materials to be between 6 $k_B T$ and 10 $k_B T$. This range seems to match well with the temperature for maximum ZT of conventional thermoelectric materials. For the case of indirect-gap materials, neither the effective masses nor the dielectric functions are strong functions of the gap [57]. A minimum band gap is required to minimize bipolar conduction of electrons and holes. When both electrons, with energies above the Fermi level, and holes, with energies below the Fermi level, contribute to electrical conduction, the asymmetry of differential conductivity is reduced, and the Seebeck coefficient becomes very small. Because for most materials the lattice contribution to thermal conductivity is higher in materials with a larger band gap, the optimum band gap is ~6–10 $k_B T$.

In the case of direct-gap materials, all the transport parameters, such as effective masses, dielectric screening, and band nonparabolicity, depend strongly upon energy gap [57]. Sofo and Mahan [55] used the expressions for the density of states and scattering times in the Boltzmann transport equation. By optimizing carrier concentration, these researchers introduced, as did Chasmar and Stratton [56], a single material parameter (B) that maximizes ZT:

$$B \propto T^{5/2} \frac{m^{*1/2} \tau}{k_L} \tag{10.2}$$

Here, τ is the electron scattering time, T the absolute temperature, m^* the effective mass, and k_L the lattice thermal conductivity. Sofo and Mahan took into account the effective mass dependency of dominant scattering mechanisms. The ionized impurity scattering time is proportional to $m^{*1/2}$, whereas the polar optical phonon scattering time scales as $m^{*-1/2}$. The acoustic phonon scattering time depends on $m^{*-3/2}$. Using Kane's k.p method [58], Sofo and Mahan argued that the electron effective mass should be directly proportional to the band gap, and thus an optimum band gap can be identified. For the case of dominant acoustic phonon or polar optical phonon scatterings, the optimum band gap is 10 $k_B T$ (due mainly to the bipolar conduction), whereas for the case of ionized impurity scattering, the B parameter increases with band gap, so there is no optimum.

The above optimization of thermoelectric material properties applies only to homogeneous semiconductors. In the case of inhomogeneous semiconductors or metamaterials (e.g. those with embedded nanoscale inclusions or solid-state thermionic devices), it is possible to overcome many of the trade-offs. For example, one can avoid the problem of bipolar conduction by selective scattering of electrons or holes. If the heterostructure band offsets in the nanocomposite are such that, e.g. electrons can move freely while holes are localized (or severely scattered), it is possible to create a high-ZT material with very small band gap or semimetallic materials [59, 60].

10.3.5 *Electrical properties of low-dimensional and 2D material thermoelectrics*

In 1993, the pioneering work by Hicks and Dresselhaus [41] renewed the interest in thermoelectrics. The key concept was that quantum confinement of electrons and holes could dramatically increase $ZT > 1$ by independently changing the power factor. Although Hicks & Dresselhaus's idea ignited intense research into nanostructured thermoelectric materials, there are still debates about the exact role that low dimensionality may play in improving thermoelectrics.

Improving the thermoelectric power factor of quantum well/wire materials is difficult for three reasons [61–63]. First, we live in a 3D world, and any quantum well structure should be imbedded in barriers. These barriers are electrically inactive, but they add to the heat loss between the hot and the cold junctions. One cannot make the barrier too thin because the tunneling between adjacent quantum wells will broaden energy levels and reduce the improvement due to the density of states. Second, the sharp features in density of states of low-dimensional nanostructures disappear quickly as soon as there is size non-uniformity in the material. The third reason is concern for interface scattering of electrons in narrow quantum structures.

Quantum dot structures have been proposed as the 0D extension of low-dimensional thermoelectrics. A theoretical study by Mahan and Sofo [64] suggested that the best thermoelectric materials have a delta function density of states. Quantum dots fit ideally into such a picture. A single quantum dot, however, is not of much interest for building into useful thermoelectric devices (but may be of scientific interest to create localized cooling on the nanoscale). There is a fundamental difference between quantum dots and quantum wires/wells. The enhanced power factor in quantum-confined 1D and 2D structures happens in the direction perpendicular to the confinement. Thus, we benefit from sharp features in the density of states, but we can still use free electron approximation with an effective mass in the direction where electric field is applied, and heat is transported. This is not really a case in a single quantum dot.

Theoretical analysis by Humphrey and Linke [65] showed that the electronic efficiency for thermoelectric cooling or power generation can approach the Carnot limit if the electron transport between the hot and the cold reservoirs happens in a single energy level under a finite temperature gradient and a finite external voltage. Despite ideal conditions, this was an important study because it showed for the first time what we need to do to approach Carnot efficiency in a thermoelectric material (this corresponds to $ZT \to \infty$). Humphrey and Linke showed that transport at a single optimized energy level corresponds to the breakdown of the Wiedemann-Franz law. The basic idea is that whenever there is a finite energy band in which electrical conduction happens, we may have counterpropagating electrical currents from the hot to the cold reservoir and from the cold to the hot reservoir. These currents do not contribute to the net electrical conduction, but they can transport energy from the hot to the cold reservoir (i.e. electronic thermal conductivity). When the electron transport in the material happens at a single energy level, it is possible to choose it at a value so that the probability of the occupation is identical in both hot and cold reservoirs. Thus, the reservoirs at different temperatures and electrochemical potentials are in an energy-specific equilibrium through the material, and there is no net current. Under such conditions and if one neglects the lattice contribution to thermal conductivity, it is possible to have thermoelectric energy generation or refrigeration with an efficiency approaching the Carnot limit. For energy-specific equilibrium, the material should be designed for a given temperature difference between hot and cold reservoirs. If the temperature difference is changed, a different material is needed.

One should note some fundamental differences between Humphrey and Linke's [65] work and that of Mahan and Sofo [64] discussed above. The latter work was based on the Boltzmann transport equation, in which the density of states is replaced by a delta function but the electronic group velocity (effective mass) is not changed. However, as we have seen above, both density of states and electron effective mass are derived from the band dispersion, and it is not possible to have delta function density of states without affecting the electron group velocity or localization. Humphrey and Linke's work, in contrast, focused on efficiency near zero output power (zero current), which is a requirement to reach the Carnot limit.

In a very nice publication, Linke and coworkers [66] studied the efficiency of quantum dot thermoelectrics in a maximum-output-power regime. In this case, the coupling between the quantum dot energy level and the reservoirs gives a fundamental broadening that reduces the efficiency of the thermoelectric energy conversion process. However, finite heat and charge currents are established, and useful work can be performed. To increase the output power, stronger couplings with reservoirs are needed. But as the coupling increases, the large broadening in the quantum well energy level reduces the energy conversion efficiency. Interestingly, these researchers obtained an optimum broadening of 2.25 $k_B T$ that maximizes the output power [66]. At the maximum output power, the efficiency of a single-level thermoelectric is only 20% of the Carnot limit.

Experimental demonstration of thermoelectric energy conversion in individual quantum dots is extremely complicated because there are often multiple energy levels inside the dot. To study transport at individual energy levels, measurements below 5 K were performed. In addition, it is very difficult to measure the temperature gradient across one quantum dot. Hoffman *et al.* [67], in a series of heroic experiments, studied thermoelectric voltage generated by a single quantum dot embedded in a nanowire. Despite parasitic heat flows and complications due to nonlinearity of the measured Seebeck voltages as a function of temperature gradient, preliminary results show electronic efficiency on the order of 88–95% of the Carnot limit for 25-nm InAs quantum dots confined by InP barriers in an InAs nanowire [68].

By trapping individual molecules between two gold electrodes with a temperature difference across them, Reddy *et al.* [69] measured junction Seebeck coefficients. The results were consistent with the calculations by Paulsson and Datta [70] and Murphy *et al.* [71]. These outstanding experiments first explored the thermoelectric properties of individual molecules. Although the single-molecule junction Seebeck measurement was a great scientific experiment, it is sometimes argued that once the thermoelectric properties of individual molecules are optimized, we can make a good bulk thermoelectric material based on layers of the same molecules. Two important issues have to be considered. First, the lattice structure is the main determinant of the electrical and thermal conductivities of most solids. Bonds, sharing of electrons with nearest neighbors, and transport of electron and phonons over several atoms dominate conductivities of good metals and semiconductors. It is hard to predict the transport coefficients of solids using only single-atom or single-molecule properties. Second, in single-molecule electrical and thermal transport measurements, the inherent properties of the electrode/molecule junction are characterized. We cannot really discuss the Seebeck coefficient of, e.g. an individual molecule, only that of a contact/molecule/contact junction. This is because electrons go ballistically across the molecule. Changing the electrodes changes the band alignments and the measured thermovoltage. The results from junction measurements are not directly related to the results obtained in solid materials made from the same molecule.

In an important paper, Kim *et al.* [72] revisited the large thermoelectric power factor in low dimensional structures, and they pointed out that the normalization of low-D electrical conductivity to the thickness or cross-section of the confinement layer is the major source of the diverging large power factors. The actual improvement in the thermoelectric power factor 'per conduction channel' is only 12–40%. Kim *et al.* also derived a minimum packing density for low-dimensional thermoelectric materials (even with ideal infinite barrier confinement) to have any improvement compared with the bulk. The importance of calculating thermoelectric power factor per conduction channel has been unfortunately ignored in many later publications that focus on thermoelectric properties of 2D materials [72]. While good power factor can be extracted if electrical conductivity is divided by the thickness of single atom layer (e.g. 0.1–1 nm), electron wavefunction has larger characteristics lengths and 2D materials cannot be packed at distances smaller than 5–10's nm. Otherwise coupling between layers will create an effective 3D material where gains from surface and 2D states are lost.

10.3.6 What can we learn from nanoelectronics and nanophotonics in energy applications?

Precise control of a material's electrical, thermal, optical, magnetic and structural properties has huge implications for thermoelectric and solid-state cooling applications. However, we should also be aware of major differences between 'information' devices and 'energy' devices. The former can be as small as possible since a 'bit' does not have a minimum size. A property of a single electron or a single photon can define a bit. On the other hand, energy applications have a 'required' cumulative length scale that matches human consumption (e.g. energy needed to keep the temperature of a house constant or energy needed to transport 1 kg of material by 100 km). This distinction is important and is one of the reasons why some of the early industrial innovations of the 20th century (e.g. airplanes) no longer exhibit exponential growth. Between 1900 and 1950, the speed of airplanes and the energy efficiency for transport increased almost exponentially whereas in the last more than 40 years, the speed of commercial airplanes has stayed constant and fuel efficiency has improved very slowly [73].

There are potential applications of 2D materials or low-dimensional thermoelectrics for localized temperature control or energy harvesting in nanoscale devices, however, it is hard to image how they can be scaled up to solve societal-scale MW/GW energy challenge and help with the climate change.

10.3.7   *Polymer/organic thermoelectrics*

Historically, thermoelectric materials have required high-temperature processing methods to make reasonably high-quality crystals. To reduce the cost, solution-processable conjugated molecules [74–76] and colloidal quantum dots [77] have recently received increased attention. However, their power factors ($S^2\sigma$) remain relatively small compared to the state-of-the-art bulk inorganic thermoelectric materials. Composites of solution-processed inorganic nanoparticles and evaporated organic thin films are limited by low electrical conductivities. In contrast, solution-processed conducting polymers have sufficiently high electrical conductivities but lack competitively high thermopowers [74]. See the excellent recent review by Russ *et al.* on organic thermoelectric materials [78]. The maximum power factors are typically obtained when the electrical conductivity is pushed to the highest, e.g. with a very large doping density, which, however, can result in a significant increase in the electronic thermal conductivity at the same time in these organic materials. The thermal conductivity of these highly-doped organic materials have not been much studied due to the difficulties in measuring the in-plane thermal conductivity of such thin films with sub-micron thicknesses.

All-organic composites composed of carbon nanotubes (CNTs) and conjugated polymers have demonstrated intriguing composite properties but a maximum ZT of only 0.02 [79]. In this case, an enhanced electrical conductivity is provided by the carbon nanotube networks formed in the composite, and a low thermal conductivity is offered by the polymer matrix. However, neither shows a particularly high Seebeck coefficient.

Recently, it is shown that the thermoelectric properties of carbon nanotube networks embedded in a non-conducting polymer matrix such as polydimethylsiloxane (PDMS) can be highly tunable and largely determined by the tunneling transport at the nanoscale junctions between CNTs [80]. The key parameters here are the junction distance and the barrier height, both of which need to be lower to simultaneously achieve a high electrical conductivity and a high Seebeck coefficient for CNT networks. Furthermore, films of semiconducting CNTs only, without metallic CNTs, have shown very large Seebeck coefficients (>400 μV/K) with reasonably high electrical conductivities [81]. Doping of CNTs further enhanced the power factor. All these recent results show that CNT-based polymer composites can be potentially a highly efficient, flexible thermoelectric materials for wearable energy harvesting and low-grade waste heat recovery.

Solution-processable polymer-inorganic hybrid materials with noticeable thermoelectric properties were recently demonstrated [82, 83]. This hybrid material combines a high-electrical-conductivity polymer with an inorganic nanoparticle with a high Seebeck coefficient. Additionally, attention was paid to work function alignment to open up the possibility of energy-dependent scattering. The result is a hybrid organic-inorganic material with a higher power factor than either the nanoparticle thin film or the polymer thin film but also with the property of polymer-like thermal transport. The measured $ZT \sim 0.1$ is promising.

10.4 Role of Advanced Computing Power

A key enabling factor for the microelectronics industry has been the emergence of computer aided design (CAD) tools which enable the design of extremely complex systems with billions of building blocks. The computing power has now increased to a point that we can do first principle calculation of many material properties. This opens up the opportunity to design new materials and predict their properties. Initiative, such as 'Materials Genome Initiative for Global Competiveness' in the US as well as machine learning for material discovery could pave the way for major applications in the energy field where, for example, one could design, thermoelectric devices with appropriate thermal and electronic transport properties from first principles and thus minimize the trial and error often used to make existing devices. An interesting example is the theoretical identification of boron arsenide (BA) as high thermal conductivity material by [84]. Based on the detailed analysis of three-phonon processes, they predicted that large separation between acoustic and optical phonon branches in BAs should have large thermal conductivity approaching diamond. A prediction before a material is synthesized is quite valuable! Following extensive efforts, two groups simultaneously demonstrated that BAs indeed have high thermal conductivity. The two groups are Tian *et al.* [85] and Kang *et al.* [86]. High measured thermal conductivities of ~ 1000 W/(m K) are respectable, but they do not quite approach thermal conductivity of diamond. It turns out that four-phonon processes start to dominate which were not included in the original prediction. This is an area where large-scale computation combined with combinatorial synthesis techniques may allow us to identify brand new complex thermoelectric energy conversion materials.

10.5 Material Cost/Efficiency Trade-Off

A key factor for studying the cost/efficiency trade-off in thermoelectric power generation is optimization of the thermoelectric module together with the heat source and the heat sink [87]. Several studies focused mostly on the thermoelectric element. However, heat flow is a significant factor affecting the output power. Through the use of an effective thermal network, the co-optimization of the thermoelectric generators and the heat sink can be performed for various heat sources. The pumping power needed for the convection heat sink is subtracted from the generated power to calculate the net output power. Yazawa and Shakouri [88] estimated the cost of material in a thermoelectric power generation system (in dollars per watt) as a function of heat source power density for different material properties (ZT) and as a function of module design (fractional area coverage of thermoelectric elements). These results provided in detail in

Chapters 5 and 6 showed that the module design plays a key factor in determining the cost of waste-heat-recovery thermoelectric systems. It is possible to bring down the material cost from \$1–2 per watt to \$0.05–0.1 per watt without improving ZT if thin-film modules with hundreds-of-micrometers-thick elements with low fractional area coverage (5–10%) and small parasitic resistances, e.g. a contact resistance in the low $10^{-6}\,\Omega\,\text{cm}$, can be developed.

The recent works in the optimization of thermoelectric design for performance and cost, including [89, 90], etc., draw a broader picture of designing thermoelectric modules in conjunction with material properties, which can be tuned in manufacturing processes and costs. The deep interrelation between the heat transfer performance of the thermal contacts and the design of the thermoelectric module and leg is gradually becoming a common knowledge of the community.

10.6 Summary

In this chapter, we reviewed recent advances in semiconductor thermoelectric physics and materials. Not only have ZT values exceeding 1.5 and higher been demonstrated reproducibly, but new concepts and enhanced understanding have emerged that will help drive the field forward in the years to come [29]. There have been questions about some of the recently reported experimental results and theoretical concepts. This is expected for a rapidly growing and dynamic field, especially because accurate thermal and thermoelectric characterization techniques for small-size samples are being developed at the same time. A major shortcoming for new researchers entering in the thermoelectric field is that very few papers clearly describe the major unsolved controversies. The Thermoelectric Handbook: Macro to Nano [91] is very helpful, as it summarizes latest advances in the field. However, as is common in books that try to capture a rapidly developing field, a few chapters describe experimental results and/or theoretical concepts that many in the community are still questioning. Unfortunately, there is no way to differentiate between the technically sound chapters and those that are speculative or even possibly wrong. Similar concerns were raised about some highly cited thermoelectric papers in top journals, e.g. Ref. [92]. The review by Vineis *et al.* [29] tries to highlight some of the open issues. It will be extremely helpful to continue healthy and open discussions in the thermoelectric community. This is how we can benefit from the talented, young researchers who try to identify and work on important energy-related problems.

Thermoelectric research is an exciting cross-disciplinary field for chemists, physicists, materials scientists, and engineers. For the field to continue its quick development, the new scientific and engineering concepts need to be coupled to new applications both in small-scale cooling and in power generation. In addition to improvements in material ZT, advances in reliable metallization and interconnects, diffusion barriers, stress management under thermal cycling, and novel module design and geometries are critical for the development of new products and markets.

There is not much work in industrial scale thermoelectric system deployment with detailed experimental and theoretical analysis of TE co-generation (especially for topping-cycle applications). Often, university researchers and funding agencies disregard system integration as this is just another engineering optimization problem. The global energy challenge for climate change is a huge problem. We won't be off fossil sources anytime soon. There are interesting trade-offs and material/device discovery if we can build prototype systems and study their performance in the field. Initial solar cell installations in 1970's or 1980's were very expensive, but we learned a lot from degradation mechanisms and field issues. This has been the basis of resurrection of photovoltaics in 2000's and 2010's. If thermoelectric systems can increase the efficiency of conventional power plants and small-scale distributed generators, there may be opportunities for MW and GW applications so we should not shy away from projects focused on system integration. Of course, this is only helpful if system integration is highly modular, different technologies could be tested and various parameters carefully studied with complementary modeling effort.

References

[1] C. Behar, J. M. Jancovici, P. Knoche, E. Dutheil, O. Lamarre, N. Devictor, K. Soederholm, *et al.* (2016). Atoms for the future 2016-Lectures, Retrieved from http://www.sfen.org/sites/default/files/public/atoms/files/aff2016_jm.jancovici.pdf.

[2] S. C. Trümper, Commercial fleets as early markets for electric vehicles, In *16th International IEEE Conference on Intelligent Transportation Systems* (ITSC 2013), 1941–1946 (2013).

[3] Honda (1999) EVPlus. Retrieved from https://global.honda/heritage/timeline/producthistory/automobiles/1997EV-Plus.html.

[4] M. Lamarche, Evaluation of the Chrysler TEVan electric vehicle. Transport Canada, transportation development centre: No. TP 12749E (1996).

[5] Electric transportation applications ford ranger EV. Retrieved from https://avt.inl.gov/sites/default/files/pdf/fsev/ford_eva.pdf.

[6] N. Melton, J. Axsen and D. Sperling, Moving beyond alternative fuel hype to decarbonize transportation, *Nature Energy*, **1**, 16013 (2016).

[7] J. M. Cullen and L. M. Allwood, The efficient use of energy: Tracing the global flow of energy from fuel to service, *Energy Policy*, **38**(1), 75–81 (2020).

[8] A. Sieminski, International energy outlook 2013. US Energy Information Administration (EIA), Report Number: DOE/EIA-0484 (2013).

[9] U.S. Energy Information Administration (EIA) Office of Energy Analysis. International energy outlook 2019 with projections to 2050, Report Number: IEO2019 (2019).

[10] V. Smil, Energy in the twentieth century: Resources, conversions, costs, uses, and consequences, *Annu. Revi. of Ener. and the Envir.*, **25**(1), 21–51 (2000).

[11] F. Katiraei, M. R. Iravani and P. W. Lehn, Micro-grid autonomous operation during and subsequent to islanding process, *IEEE Trans. on pow. deli.*, **20**(1), 248–257 (2005).

[12] E. Apostolaki-Iosifidou, P. Codani and W. Kempton, Measurement of power loss during electric vehicle charging and discharging energy, **127**, 730–742 (2017).

[13] All-Electric Vehicles, US Department of energy office of energy efficiency & renewable energy, Retrieved from https://www.fueleconomy.gov/feg/evtech.shtml.

[14] L. Bell, Cooling, heating, generating power, and recovering waste heat with thermoelectric systems, *Science*, **321**(5895), 1457–1461 (2008).

[15] M. Kishi, H. Nemoto, T. Hamao, M. Yamamoto, S. Sudou, M. Mandai and S. Yamamoto, Micro thermoelectric modules and their application to wristwatches as an energy source, *In Eighteenth International Conference on Thermoelectrics. Proceedings, ICT'99* (Cat. No. 99th 8407), 301–307 (1999).

[16] M. Kawata and A. Takakura, Thermoelectrically powered wrist watch. U.S. Patent, **5**(889), 735 (1999).

[17] K. Yazawa and A. Shakouri, Thermoelectric topping cycles with scalable design and temperature dependent material properties, *Scripta Materialia*, **111**, 58–63 (2016).

[18] H. Ritchie, Number of people in the world without electricity falls below one billion, Our World in Data (2019). Retrieved from https://ourworldinda ta.org/number-of-people-in-the-world-without-electricity-access-falls-below-one-billion.

[19] Least Developed Countries (LDCs), Department of economic and social affairs, United Nation, Retrieved from https://www.un.org/development/desa/dpad/least-developed-country-category/ldc-data-retrieval.html.

[20] Retrieved from https://www.bioliteenergy.com/products/firepit.

[21] D. Champier, J. P. Bedecarrats, R. Rivaletto and F. Strub, Thermoelectric power generation from biomass cook stoves, *Energy*, **35**(2), 935–942 (2010).

[22] World Health Organization, Household air pollution and health, (2018), Retrieved from https://www.who.int/news-room/fact-sheets/detail/house hold-air-pollution-and-health.

[23] M. B. Blarke, K. Yazawa, A. Shakouri, and C. Carmo, Thermal battery with CO_2 compression heat pump: Techno-economic optimization of a high-efficiency Smart Grid option for buildings, *Ener. and Build.*, **50**, 128–138 (2012).

[24] T. Wang, O. Kurtulus, K. Yazawa, and E. Groll, Experimental study of a CO_2 thermal battery for simultaneous cooling and heating use, *Proceedings of the 15th International Refrigeration and Air Conditioning Conference at Purdue*, 2701 (2014).

[25] K. Yazawa, V. Wong, M. B. Blarke and A. Shakouri, Modeling and experiments on high response thermo-energy buffer for intermittent sources, *Proceedings of ASME International Mechanical Engineering Congress and Exposition*, IMECE 2012–86623 (2012).

[26] A. Shakouri, Recent developments in semiconductor thermoelectric physics and materials, *Annu. Rev. Mater. Res.*, **41**, 399–431 (2011).

[27] G. J. Snyder and E. S. Toberer, Complex thermoelectric materials, *Nat. Mater.*, **7**, 105–114 (2008).

[28] J. R. Sootsman, D. Y. Chung and M. G. Kanatzidis, New and old concepts in thermoelectric materials, *Angew. Chem. Int. Ed.*, **48**(46), 8616–39 (2009).

[29] C. J. Vineis, A. Shakouri, A. Majumdar and M. G. Kanatzidis, Nanostructured thermoelectrics: Big efficiency gains from small features, *Adv. Mater.*, **22**(36), 3970–80 (2010).

[30] G. Tan, L. D. Zhao and M. G. Kanatzidis, Rationally designing high-performance bulk thermoelectric materials, *Chem. Rev.*, **116**, 12123–12149 (2016).

[31] J. He and T. M. Tritt, Advances in thermoelectric materials research: Looking back and moving forward, *Science*, **357**(6358), eaak9997 (2017).

[32] D. M. Rowe, ed., Handbook of thermoelectrics, *Boca Raton*, CRC (1995).

[33] A. Shakouri and M. Zeberjadi, Nanoengineered materials for thermoelectric energy conversion, In *Thermal Nanosystems and Nanomaterials, ed. S.*, **2**, 225–99. Berlin/Heidelberg: Springer (2009).

[34] T. E. Humphrey, M. F. O'Dwyer and H. Linke, Power optimization in thermionic devices, *J. Phys. D*, **38**, 2051–54 (2005).

[35] M. Zebarjadi, Solid-State thermionic power generators: An analytical analysis in the nonlinear regime, *Phys. Rev. Appl.*, **8**, 014008 (2017).

[36] J. H. Bahk, Z. Bian and A. Shakouri, Electron energy filtering by a nonplanar barrier to enhance the thermoelectric power factor in bulk materials, *Phys. Rev. B*, **87**, 075204 (2013).

[37] A. Shakouri, Nanoscale thermal transport and microrefrigerators on a chip, *Proce. of the IEEE*, **94**(8), 1613–1638 (2006).

[38] A. F. Ioffe, Semiconductor thermoelements and thermoelectric cooling, London: Infosearch (1957).

[39] G. S. Nolas, J. W. Sharp and H. J. Goldsmid, Thermoelectrics: Basic principles and new materials developments. Berlin/Heidelberg: Springer (2001).

[40] M. S. Dresselhaus, G. Dresselhaus, X. Sun, Z. Zhang, S. B. Cronin and T. Koga, Low-dimensional thermoelectric materials, *Phys. Sol. Sta.*, **41**, 679–682 (1999).

[41] L. D. Hicks and M. S Dresselhaus, Effect of quantum-well structures on the thermoelectric figure of merit, *Phys. Rev. B*, **47**, 12727–31 (1993).

[42] M. S. Dresselhaus, Y. M. Lin, S. B. Cronin, O. Rabin, M. R. Black, *et al.* Quantum wells and quantum68. X. F. Fan, G. H. Zeng, C. LaBounty, J. E. Bowers, E. Croke, *et al.* SiGeC/Si superlattice microcoolers, *Appl. Phys. Lett.*, **78**, 1580–82 (2001).

[43] A. Shakouri and J. E. Bowers, Heterostructure integrated thermionic coolers, *Appl. Phys. Lett.*, **71**, 1234–36 (1997).

[44] G. D. Mahan, Thermionic refrigeration, *Semicond. Semimet.*, **71**, 157–74 (2001).

[45] R. Venkatasubramanian, Phonon blocking electron transmitting superlattice structures as advanced thin film thermoelectric materials, *Semicond. Semimet.*, **71**, 175–201 (2001).

[46] G. Chen, Phonon transport in low-dimensional structures, *Semicond. Semimet.*, **71**, 203–59 (2001).

[47] M. Zebarjad, K. Esfarjani, A. Shakouri, J. H. BahkZ. X. Bian, *et al.*, Effect of nano-particle scattering on thermoelectric power factor, *Appl. Phys. Lett.*, **94**, 202105 (2009).

[48] J. P. Heremans, V. Jovovic, E. S. Toberer, A. Saramat, K. Kurosaki, A. Charoenphakdee, S. Yamanaka and G. J. Snyder, Enhancement of thermoelectric efficiency in PbTe by distortion of the electronic density of states, *Science*, **321**, 554–557 (2008).

[49] Y. Pei, X. Shi, A. LaLonde, H. Wang, L. Chen and G. J. Snyder, Convergence of electronic bands for high performance bulk thermoelectrics, *Nature*, **473**, 66–69 (2011).

[50] A. Shakouri, C. Labounty, P. Abraham, J. Piprek and J. E. Bowers, Enhanced thermionic emission cooling in high barrier superlattice heterostructures, *Mater. Res. Soc. Symp. Proc.*, **545**, 449–58. Warrendale, PA, (1998).

[51] D. Vashaee and A. Shakouri, Conservation of lateral momentum in heterostructure integrated thermionic coolers, *Mater. Res. Soc. Symp. Proc.*, **691**, 131–45. Warrendale, PA (2001).

[52] A. Shakouri, Nanoscale devices for solid state refrigeration and power generation, *Annu. IEEE Semicond. Therm. Meas. Manag. Symp., 20th*, Piscataway, NJ: IEEE, 1–9 (2004).

[53] J. H. Bahk, R. B. Prost, K. Margatan and A. Shakouri, Linerized boltzmann transport calculator for thermoelectric materials. Published on nanoHUB.org. [Online]. https://nanohub.org/tools/btesolver.

[54] N. W. Ashcroft and N. D. Mermin, Solid state physics. New York: Holt, Rinehart, and Winston (1976).

[55] J. O. Sofo and G. D. Mahan, Optimum band gap of a thermoelectric material, *Phys. Rev. B*, **49**, 4565–70 (1994).

[56] R. P. Chasmar and R. Stratton, The thermoelectric figure of merit and its relation to thermoelectric generators, *J. Electron. Control*, **7**(1), 52–72 (1959).

[57] G. D. Mahan, Figure of merit for thermoelectrics, *J. Appl. Phys.*, **65**, 1578–83 (1989).

[58] E. O. Kane, The k.p method. In semiconductors and semimetals, **1**, ed. R. K. Willardson, AC Beer, **3**, 75–100, Amsterdam: Elsevier (1966).

[59] J. H. Bahk and A. Shakouri, Minority carrier blocking to enhance the thermoelectric figure of merit in narrow band gap semiconductors, *Phys. Rev. B*, **93**, 165209 (2016).

[60] M. Markov, X. Hu, H. C. Liu, N. Liu, S. J. Poon, K. Esfarjani and M. Zebarjadi, Semi-metals as potential thermoelectric materials, *Sci. Rep.*, **8**, 9876 (2018).

[61] D. A. Broido and T. L. Reinecke, Effect of superlattice structure on the thermoelectric figure of merit, *Phys. Rev. B*, **51**, 13797–800 (1995).

[62] J. O. Sofo and G. D. Mahan, Thermoelectric figure of merit of superlattices, *Appl. Phys. Lett.*, **65**, 2690–92 (1994).

[63] G. Chen and A. Shakouri, Heat transfer in nanostructures for solid-state energy conversion, *J. Heat Transf. Trans. ASME*, **124**(2), 242–52 (2002).

[64] G. D. Mahan and J. O. Sofo, The best thermoelectric, *Proc. Natl. Acad. Sci. USA*, **93**, 7436–39 (1996).

[65] T. E. Humphrey and H. Linke, Reversible thermoelectric nanomaterials, *Phys. Rev. Lett.*, **94**(9), 0966011 (2005).

[66] N. Nakpathomkun, H. Q. Xu and H. Linke, Thermoelectric efficiency at maximum power in lowdimensional systems, *Phys. Rev. B*, **82**, 235428 (2010).

[67] E. A. Hoffmann, H. A. Nilsson, J. E. Matthews, N. Nakpathomkun, A. I. Persson, *et al.*, Measuring temperature gradients over nanometer length scales, *Nano Lett.*, **9**, 779–83 (2009).

[68] E. Hoffmann, The thermoelectric efficiency of quantum dots in indium arsenide/indium phosphide nanowires. PhD thesis. Univ. Or., (2009).

[69] P. Reddy, S. Y. Jang, R. A. Segalman and A. Majumdar, Thermoelectricity in molecular junctions, *Science*, **315**, 1568–71 (2007).

[70] M. Paulsson and S. Datta, Thermoelectric effect in molecular electronics, *Phys. Rev. B*, **67**, 241403 (2003).

[71] P. Murphy, S. Mukerjee and J. Moore, Optimal thermoelectric figure of merit of a molecular junction, *Phys. Rev. B*, **78**(16), 161406 (2008).

[72] R. Kim, S., Datta and M. S. Lundstrom, Influence of dimensionality on thermoelectric device performance, *J. Appl. Phys.*, **105**, 034506 (2009).

[73] https://www.mercatus.org/publications/technology-and-innovation/ airplane-speeds-have-stagnated-40-years.

[74] A. Shakouri and S. Li, Thermoelectric power factor for electrically conductive polymers, *Int. Conf. Thermoelectr.*, 18th, Piscataway, NJ: IEEE, 402–6 (1999).

[75] P. Reddy, S. Y. Jang, R. A. Segalman and A. Majumdar, Thermoelectricity in molecular junctions, *Science*, **315**, 1568–71 (2007).

[76] A. Casian, Thermoelectric properties of electrically conducting organic materials, *See Ref.*, **147**(36), 1–8 (2006).

[77] M. V. Kovalenko, B. Spokoyny, J.-S. Lee, M. Scheele and A. Weber, Semiconductor nanocrystals functionalized with antimony telluride zintlions for nanostructured thermoelectrics, *J. Am. Chem. Soc.*, (2010).

[78] B. Russ, A. Glaudell, J. J. Urban, M. L. Chabinyc and R. A. Segalman, Organic thermoelectric materials for energy harvesting and temperature control, *Nat. Rev. Mater.*, **1**, 16050 (2016).

[79] J. L. Blackburn, A. J. Ferguson, C. Cho and J. C. Grunlan, Carbon-Nanotube-based thermoelectric materials and devices, *Adv. Mater.*, **30**, 1704386 (2018).

[80] R. Prabhakar, M. S. Hossain, W. Zheng, P. K. Athikam, Y. Zhang, Y. Y. Hsieh, E. Skafidas, Y. Wu, V. Shanovand J. H. Bahk, Tunneling-Limited thermoelectric transport in carbon nanotube networks embedded in Poly(dimethylsiloxane) elastomer. *ACS Appl. Ener. Mater.*, **2**, 2419–2426 (2019).

[81] A. D. Avery, B, H. Zhou, J. Lee, E. Lee, E. M. Miller,R. Ihly, D. Wesenberg, K. S. Mistry, S. L. Guillot, B. L. Zink, *et al.*, Tailored semiconducting carbon nanotube networks with enhanced thermoelectric properties, *Nat. Energy*, **1**, 16033 (2016).

[82] K. C. See, J. P. Feser, C. E. Chen, A. Majumdar, J. J. Urban and R. A. Segalman, Water-processable polymernanocrystal hybrids for thermoelectrics, *Nano Lett.*, **10**(11), 4664–67 (2010).

[83] J. H. Bahk, H. Fang, K. Yazawa and A. Shakouri, Flexible thermoelectric materials and device optimization for wearable energy harvesting. *J. Mater. Chem. C*, **3**, 10362–10374 (2015).

[84] L. Lindsay, D. A. Broido and T. L. Reinecke, First-principles determination of ultrahigh thermal conductivity of boron arsenide: A competitor for diamond?, *Phys. Rev. Lett.*, **111**(2), 025901 (2013).

[85] F. Tian, B. Song, X. Chen, N. K. Ravichandran, Y. Lv, K. Chen, S. Sullivan, *et al.*, Unusual high thermal conductivity in boron arsenide bulk crystals, *Science*, **361**(6402), 582–585 (2018).

[86] J. S. Kang, M. Li, H. Wu, H. Nguyen and Y. Hu, Experimental observation of high thermal conductivity in boron arsenide, *Science*, **361**(6402), 575–578 (2018).

[87] K. Yazawa and A. Shakouri, Efficiency of thermoelectric generators at maximum output power, *Proc. Int. Conf. Thermoelectr., Shanghai. Piscataway, NJ: IEEE* (2010).

[88] K. Yazawa and A. Shakouri, Energy payback optimization of thermoelectric power generator systems, *Proc. ASME Int. Mech. Eng. Congr. Expo.* (IMECE2010), Vancouver: IMECE2010–37957 (2010).

[89] S. LeBlanc, S. K. Yee, M. L. Scullin, C. Dames and K. E. Goodson, Material and manufacturing cost considerations for thermoelectrics, *Ren. and Sust. Ener. Rev.*, **32**, 313–327 (2014).

[90] T. J. Hendricks, S. K. Yee and S. LeBlanc, Cost scaling of a real-world exhaust waste heat recovery thermoelectric generator: A deeper dive, *J. of Elec. Mat.*, **45**(3), 1751–1761 (2016).

[91] D. M. Rowe, ed., Thermoelectric handbook macro to nano, Boca Raton, FL: CRC (2006).

[92] T. C. Harman, P. J. Taylor, M. P. Walsh and B. E. LaForge, Quantum dot superlattice thermoelectric materials and devices, *Science*, **297**(5590), 2229–32 (2002).

Nomenclature

A	area, cross-section area, m^2
a	length, m
b	width, m
C	heat capacitance, J/K
	solar concentration ratio, dimensionless
	material market price (cost), \$/kg
c	pitch, m
C_l	elastic constant, N/m^2
C_m	material cost of module, \$/m^2
COP	coefficient of performance, dimensionless
C_p	specific heat, J/(kg K)
D	distance, m
d	thermoelectric leg length, m
	thickness, m
d_0	optimum leg length, m
D_a	acoustic phonon deformation potential, eV
D_h	hydraulic diameter, m
E	energy, J or eV
e	elementary charge, C
E_F	Fermi energy, eV
E_g	energy level, J or eV
E_p	specific exergy rate, J/(m^2 s)
E_x	exergy, J/m^2
F	fill factor, dimensionless

f	frequency, Hz
	solar energy density ratio, dimensionless
f_0	Fermi−Dirac distribution function, dimensionless
F_r	geometrical factor, dimensionless
G	mass flow rate, kg/s
	unit price, \$/kg
g	gravity, m/s^2
H	hours of operation, hour, height, m
h	heat transfer coefficient, W/(m^2 K)
	thickness of substrate, m
I	electrical current, A
	initial cost, \$/(W m^2)
J	electric current density, A/m^2
K	Interfacial shear compliance, 1/(Pa m)
	thermal conductance, W/K
k	shear stress parameter, 1/m
k_B	Stefan–Boltzmann constant, W/(m^2 K^4)
L	characteristic length, m
	distance between two legs in a module, m
l	half width of leg, m
M	Total mass per unit area, kg/m^2
m	electrical resistance ratio, dimensionless
m^*	effective mass of electron, kg
N	number of channels
	number of legs
N_v	non-ionized defect density, 1/m^3
P	pressure, Pa
p	pitch, m
Q	heat, W
q	heat flux or heat current density, W/m^2
R	electrical resistance, Ω
r_0	static screening length, m
S	Seebeck coefficient, V/K
s	entropy, J/K
	randomness of alloy, dimensionless
$\dot{S}_{\text{gen}}$	entropy generation rate, J/(K s)

T	temperature, K
t	time, s
$\bar{T}$	mean temperature, K
U	cost per power, \$/W, heat transfer coefficient, $W/(m^2\,K)$
u	velocity, m/s
U_a	alloy scattering potential, eV
U_v	short-range potential of defect, eV
V	voltage, V
$\dot{V}$	volumetric flow rate, m^3/s
W	power, W
	characteristic width, m
w	power per unit area, W/m^2
X	transformational thermal resistance, K/W
x	relative leg length to optimum $= (d/d_0)$, dimensionless
x	distance, m
Y	energy cost, \$/Wh
	transformational thermal resistance, K/W
Z	figure of merit of thermoelectric material, 1/K
	impedance, Ω
ZT	figure of merit, dimensionless

Greek symbols

α	coefficient of thermal expansion, 1/K
	nonparabolicity, 1/eV
	thermal diffusivity, m^2/s
	thermal resistance ratio, dimensionless
β	thermal conductivity, $W/(m\,K)$
β_{TCR}	temperature coefficient of resistance, 1/K
ε	effectiveness, dimensionless
	permittivity, F/m
γ	ratio, dimensionless
η	efficiency, dimensionless
ϕ	phase delay, radian
κ	interfacial shear compliance, m/Pa
	thermal conductivity, $W/(m\,K)$
	thermal resistance ratio, dimensionless

λ	eigen value, dimensionless
μ	Thomson coefficient, V/K
	viscosity, Pa s
ν	carrier velocity, m/s
	Poisson ratio, dimensionless
Ω	primitive cell volume, m^3
ρ	density, kg/m^3
	electrical resistivity, Ω m
ρ_{DOS}	electric density of state, dimensionless
σ	electrical conductivity, $1/(\Omega\, m)$
	Stefan–Boltzmann constant, $W/(m^2\, K^4)$
τ	time constant, s
	relaxation time of carrier, s
ω	angular frequency, radian/s
Ξ	exergy per unit area, W/m^2
ψ	thermal resistance, K/W, $K\, m^2/W$

Subscripts

0	optimum (at maximum power output)
a	air
	ambient
Abs	absorber
b	surrounding of earth
bi	bipolar
c	cold side
ch	channel
e	electrons
	electronic contribution
	thermoelement
eff	effective
f	fluid, fuel
g	interfacial mid-point
h	heat source side
	holes (positive charge carriers)
HS	heat sink

in	input
J	Joule effect component
j	junction
L	lattice contribution
	load
m	intermediate point
	module
max	maximum
oc	open circuit
opt	optimum (at maximum power output)
out	output
P	Peltier effect component
p	phonons
	planet (earth)
pair	pair of n-type and p-type
s	heat source
	substrate
	solar
ST	steam turbine
sub	substrate
TE	thermoelectric
TF	thin film
th	thermal
u	per unit mass
w	thermoelectric thread
wett	wetted section
y	warp thread (insulator)

Index

Printed in the United States
by Baker & Taylor Publisher Services